Daniel Andrade Miranda

Archaean geological evolution in the extreme south of the São Francisco Craton

Daniel Andrade Miranda

Archaean geological evolution in the extreme south of the São Francisco Craton

Serra da Babilônia region, Bom Sucesso - MG

ScienciaScripts

Imprint

Cover image: www.ingimage.com

This book is a translation from the original published under ISBN 978-3-330-76545-0.

Publisher:
Sciencia Scripts
is a trademark of
Dodo Books Indian Ocean Ltd. and OmniScriptum S.R.L publishing group

120 High Road, East Finchley, London, N2 9ED, United Kingdom
Str. Armeneasca 28/1, office 1, Chisinau MD-2012, Republic of Moldova, Europe
Managing Directors: Ieva Konstantinova, Victoria Ursu
info@omniscriptum.com

Printed at: see last page
ISBN: 978-620-8-51904-9

ACKNOWLEDGEMENTS

I would like to thank the Federal University of Minas Gerais and the Department of Geology for the opportunity and the studies.

To CNPq for the scholarship that contributed to the completion of this dissertation.

To my family and relatives who have always encouraged and supported me in this stage of my life. My grandmothers, Zina and Rita, who pray for me every day.

To my supervisor Alexandre de Oliveira Chaves, for his teaching, dedication and patience in helping me to complete this dissertation.

To Maritza for her unconditional support and companionship for more than 10 years.

To the friends who have welcomed me in Belo Horizonte over the last two years.

To my friends in Geology, especially the postgraduate group, who helped make my master's degree more fun.

To the teachers on the course, for their teachings and past experiences that have contributed to my professional and personal development.

Finally, to all those who contributed directly or indirectly to this work, my sincere thanks.

"The information we have about the history of the universe comes from two sources: light and rocks. *One crosses space, the other time."*

(Claude J. Allègre)

SUMMARY

This work deals with the Neoarchean magmatic and metamorphic evolution of the extreme south of the São Francisco Craton (CSF) in the Serra da Babilónia region (Bom Sucesso - MG) and is presented in the form of 2 scientific articles. The first article presents the petrographic and lithochemical characteristics of the rocks found around the Serra da Babilónia, seeking to discuss their geological evolution in relation to the Rio das Velhas Orogeny. In the southern portion of the CSF, west of the town of Bom Sucesso (MG), there are metamorphosed Neoarchean lithotypes that vary from ultramafic to mafic, enderbitic to chamockitic, and even granitic. Five distinct geological units were identified: (I) a gneiss banded with muscovite and biotite, found in the southeast of the area, is apparently related to the oldest rocks in the region, with an age of around 3000 Ma; (II) a metagranitoid called Capela São Francisco, still undated in the study area, is related to granitoids from 2750 to 2700 Ma, rich in K and of crustal origin, widely found in the southern part of the CSF. It is peraluminous, has a high silica and K content, low Ca, and shows fractionation of light rare earth elements (ETR) and negative Ce and Eu anomalies. This unit derives from the partial fusion of the paraderived crust represented by the gneisses of unit I. The presence, albeit rare, of cordierite in this metagranitoid suggests that the anatexis occurred under granulite facies conditions; (III) metachamockites to metachamoenderbites called Rio do Amparo-Bom Sucesso (27182708 Ma), derived from a mantle and crustal source, are metalluminous, calcium-alkaline and show similar fractionation of light and heavy ETR, with a negative Eu anomaly; (IV) metaenderbites to metachamoenderbites called Babylon (2730-2660 Ma) are metalluminous, rich in Na, vary from calcium to calcium-alkaline, with a fractionated ETR pattern, without Eu anomalies. Their origin is linked to the partial fusion of K-poor mafic-ultramafic mantle sources. The presence of orthopyroxene substituted for amphibole in the Rio do Amparo-Bom Sucesso and Babilónia units suggests that the crystallisation/recrystallisation of these rocks took place under a granulite facies, with retrometamorphism into an amphibolite facies; (V) metanorites/metagabronorites (2660 Ma) and peridotitic metaultramafics, the former being the product of magmatic differentiation of the latter. They are similar in their ETR patterns and spatial arrangement within the study area. Despite being recrystallised, locally they retain preserved igneous textures. Three generations of non-archean mafic dykes cut these units. This association of units records the Neoarchean evolution of an important portion of the southern CSF during the late-collisional phase of the Rio das Velhas event and is similar to those of the same age and tectonic context described in cratons on other continents.

The second article discusses U-Th-Pb chemical dating by electron microprobe carried out on monazite crystals with phosphate-silicate reaction crowns found in a metachamoenderite from the Babylon unit and the relationship between the age found and the presence of the crowns on the crystals. Coming from the Neoarchean associations of the study area, a Babylonian metachamoenderite crystallised and metamorphised into a granulite facies around 2700 Ma contains monazite crystals with apatite, allanite and epidote crowns produced by the interaction of monazite with late metamorphic fluids containing Ca, Fe, Si and Al. Using electron microprobe, the compositions of the monazite crystals were determined and their U, Th and Pb contents allowed an average Neoarchean chemical age of 2657 ± 36 Ma to be obtained. It is suggested that this

corresponds to the age of the amphibolite facies retrometamorphism of the metachamoenderbito developed during the late-collisional Neoarchean period of the Rio das Velhas event, when the phosphate-silicate reaction crowns would have developed contemporaneously with the growth of amphibole and biotite in this rock. The presentation was thus established according to the theme of each article and the results of both are necessary for the final considerations on the subject proposed here.
Key words: Neoarchean, Southern São Francisco Craton, magmatism, metamorphism, geological evolution.

SUMMARY

CHAPTER 1

PETROGRAPHY AND GEOCHEMISTRY OF NEOARCHEAN LITHOTYPES IN THE SERRA DA BABILÓNIA REGION (BOM SUCESSO - MG): GEOTECTONIC IMPLICATIONS

Daniel Andrade MIRANDA[1], Alexandre de Oliveira CHAVES[1]
It will be published in the journal Geosciences = Geosciences, Rio Claro, SP, Brazil - eISSN: 1980-900X

INTRODUCTION

The end of the Archaean (3000-2500 Ma) was a period of fundamental changes in many aspects of the geological record. According to Windley (1995), Archaean cratons are normally composed of three geological units: (1) gneissic basement of tonalitic-trondhjemitic-granodioritic (TTG) composition; (2) *greenstone belts*, and also *sill-type* ultramafic plutonic complexes (Marshall & Schonwandt, 1999); (3) intrusive granitoids.

The Neoarchean is marked by the generation of granitoid/chamockitoid rocks, which, according to Laurent et *al.* (2014), can be classified into four groups: (1) Large bodies of tonalites, trondhjemites and granodiorites (TTGs), originating from the partial fusion of metaultramafic rocks; (2) Monzodiorites and granodiorites rich in Mg, Fe and K, with metalluminous characteristics, called sanukitoids, derived from the mixture between the peridotitic mantle and a component rich in incompatible elements; (3) Biotite granites and two-mica granitoids, peraluminous, formed from the fusion of older crustal lithologies (TTGs and metasediments, respectively); and (4) Hybrid mixtures of granitoids with high-K that have characteristics of the first three groups.

Considering previous studies on the diversity of Archaean granitoids/chamockitoids in cratonic areas (Almeida et al., 2011, 2013; Dey et al., 2014) and also studies on their geotectonic evolution (Percival et al., 2006; Smithies et al., 2007; van Kranendonk et al., 2007), Laurent et *al.* (2014) realised that the same sequence is systematically repeated in several Archaean terrains (Figure 1), one stage with a long-lasting period of TTG magmatism (Stage 1), and, subsequently, another, with a much shorter duration, where the other granitoid groups form together or in rapid succession (Stage 2).

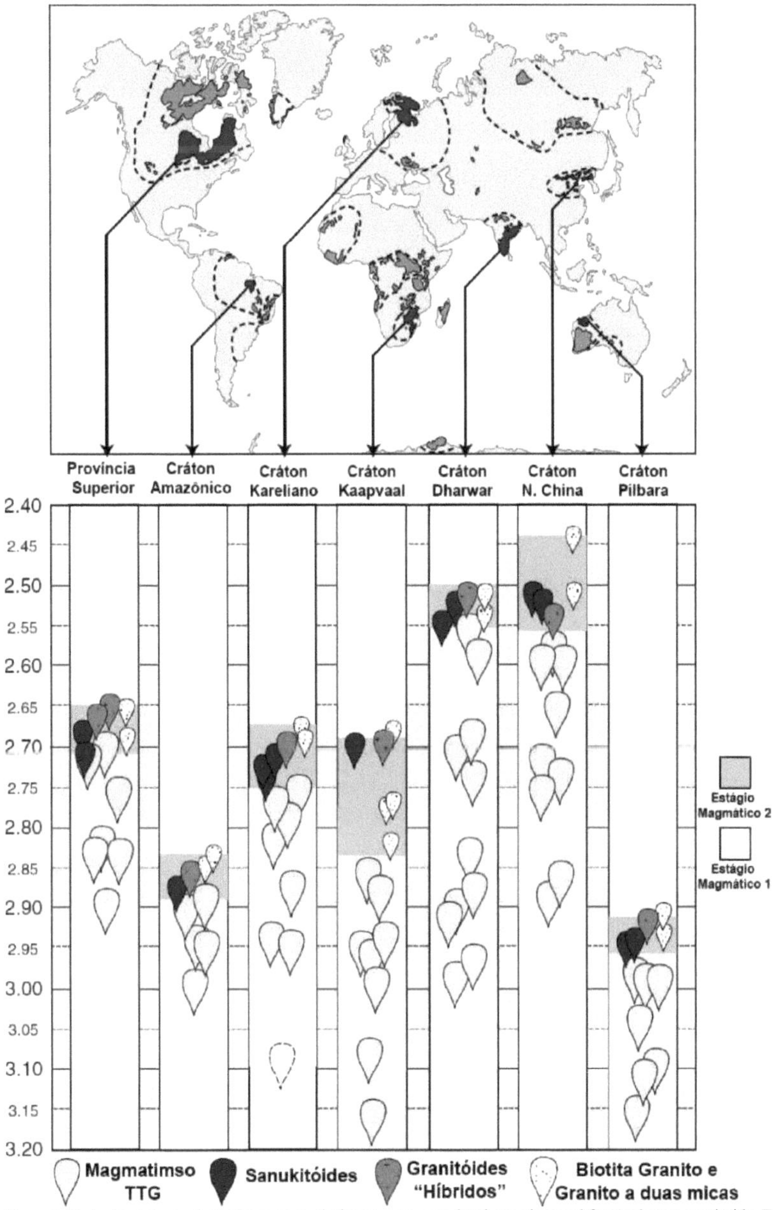

Figure 1. Cratonic regions where the same evolutionary sequence has been observed for Archaean granitoids. Extracted from Laurent *et al.* (2014).

In the southern portion of the São Francisco Craton, there are many bodies of chamockitic and enderbitic gneisses (Jordt-Evangelista, 1997; Quéméneur et al., 2003; Paciullo et al., 2007; Campos et al., 2003), as well as several bodies of K-rich metagranitoids (Carneiro, 1992; Noce, 1995; Teixeira et *al.,* 1996;

Machado et al., 1996; Romano et al., 2013) and metamorphosed mafic-ultramafic complexes (Engler et al., 2002; Campos et al., 2003; Quéméneur et al., 2003; Paciullo et al., 2007) have been identified as having been formed between 3200 Ma and 2600 Ma. All these lithotypes are found around the Serra da Babilónia (Bom Sucesso - MG) and their petrographic and geochemical aspects are discussed here within the perspective of the geodynamic evolution of Neoarchean cratonic terrains suggested by Laurent et al. (2014).

GEOLOGICAL FRAMEWORK

The southern portion of the São Francisco Craton (CSFM) is made up of granite-gneiss-migmatite terrains and Archaean greenstone belts (Rio das Velhas Supergroup), as well as Palaeoproterozoic clastic-chemical metasedimentary rocks (including itabirites in the Iron Quadrangle) of the Minas Supergroup and Neoproterozoic pelite-carbonate sedimentary covers of the Bambuí Group (Figure 2). In these Archaean terrains there are three main metamorphic complexes, called Belo Horizonte, Bonfim and Campo Belo. Most of them consist of gneisses and migmatites of tonalite-trondhjemite-granodiorite (TTG) composition, whose protoliths were formed between 3200 and 2900 Ma. Calcium-alkaline syntectonic granitoids formed in the Rio das Velhas Orogeny, with ages around 2780 Ma, post-tectonic granitoids formed between 2700 and 2600 Ma, as well as mafic and mafic-ultramafic intrusions (Noce et al., 1998; Teixeira et al., 2000), are found in the middle of the TTG terrains.

In regional terms, the evolution of the South Minas Gerais Palaeoproterozoic Province (Teixeira et al., 2005), a domain of the action of metamorphism and deformation on the archaean rocks of the CSFM, should be highlighted. It includes the Mineiro Belt at the extreme SE edge of the CSFM, a marker of Palaeoproterozoic events of an accretionary and ensialic nature. This belt, with a general SW-NE direction and a SE to NW tectonic gradient, holds the initial record of an intra-oceanic island arc from around 2300 Ma, added to an Andean-type continental margin. A magmatic arc would have been built on this margin, with plutons from 2200 Ma, until the moment of the continental collision at 2100 Ma, when the sediments of the Minas Supergroup were deformed in the Quadrilátero Ferrífero region (Teixeira & Figueiredo, 1991; Machado et al., 1992; Ávila et al., 2010; Noce et al., 2000; Alkmim, 2004).

According to Carneiro & Barbosa (2008), the sialic crustal segment of the CSFM, in which the study area is located, is made up predominantly of gneisses, granitoids, amphibolites, mafic and ultramafic rocks and chamockites, which were initially grouped by Machado Filho *et al.* (1983) into the Divinópolis and Barbacena metamorphic complexes. However, Teixeira et *al.* (1996) combined the Divinópolis and Barbacena metamorphic complexes into a single complex, which they called the Campo Belo Metamorphic Complex (CMCB), whose geological evolution began in the Mesoarchean. Supracrustal relicts are found locally, which may correspond to Archaean sequences and/or Palaeoproterozoic sequences correlating to the Iron Quadrangle (Teixeira et al., 2000). With regard to metamorphism, most of the rocks in the CMCB were generated or subjected to metamorphic conditions that reached the amphibolite and granulite facies (Femandes & Carneiro, 2000).

This whole complex is cut by mafic dykes (diabase and amphibolite) of various generations, as described by Chaves (2013).

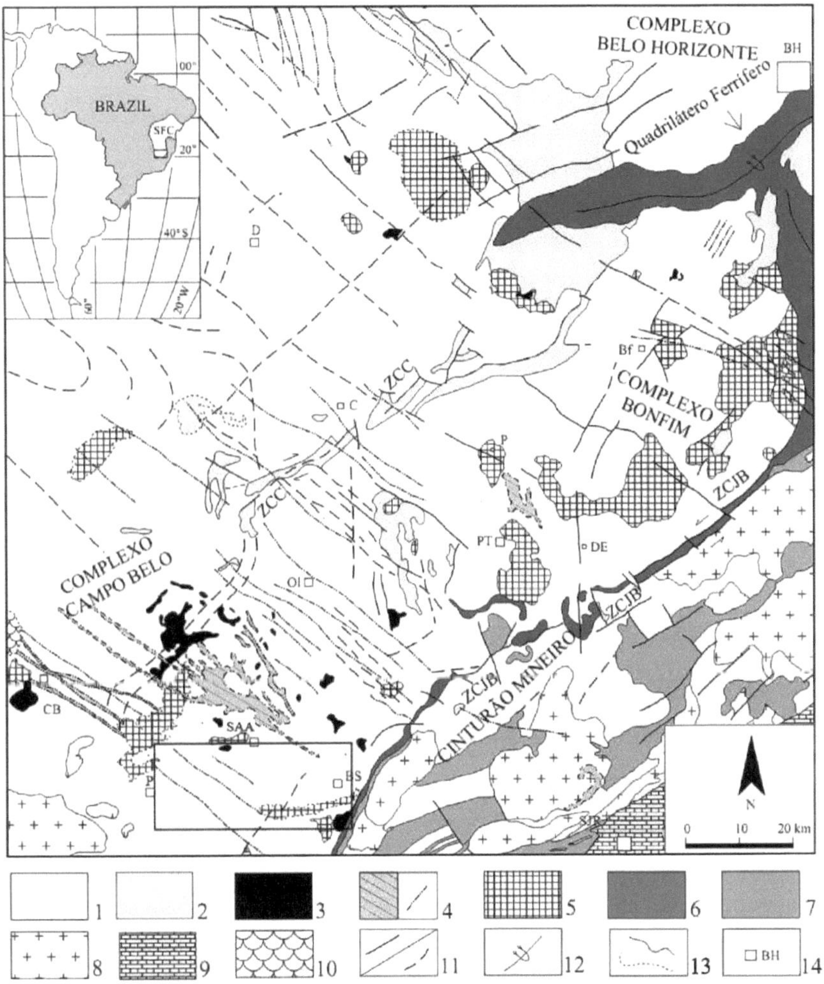

Figura 2. Schematic geological map of the southern portion of the São Francisco Craton showing the main Archaean and Palaeoproterozoic plutonic bodies (Extracted from Campos *et al.*, 2003). 1 - Archaean basement partially reworked in the Palaeoproterozoic. - Rio das Velhas *Greenstone*. 3 - Neoarchean plutonic ultramafic suites. - Neoarchean to Mesoproterozoic gabbros and diorites. 5 - Neoarchean granitoids. 6 - Minas Supergroup. 7 - Indiscriminate *greenstone* sequences. 8 - Palaeoproterozoic granitoids of the Minas Belt. 9 - São João Del Rey/Andrelândia Groups. 10 - Undivided Neoproterozoic cratonic cover - Bambuí Group. 11 - Faults and fractures (ZCC = Cláudio Shear Zone; ZCJB = Jeceaba-Bom Sucesso Shear Zone). 12 - Axial planes. 13 - Lithological contacts. 14 - Towns: BH = Belo Horizonte, Bf = Bonfim, BS = Bom Sucesso, CB = Campo Belo, C = Cláudio, DE = Desterro de Entre Rios, D = Divinópolis, O1 = Oliveira, P = Perdões, PT = Passa Tempo, SAA = Santo Antônio do Amparo, SJR = São João Del Rei. The black rectangle delimits the study area.

The study area, to the west of the Serra de Bom Sucesso, represents an Archaean terrain made up of TTG gneisses with muscovite and biotite, metachamockites to metachamoenderbites, with intercalations of ultramafic and mafic rocks, as well as potassic metagranitoids (Quéméneur *et al.,* 2003). Based on the aerogeophysical data available (Lasa Engenharia e Prospecções, 2001), the geological map in figure 3 for the area is suggested here.

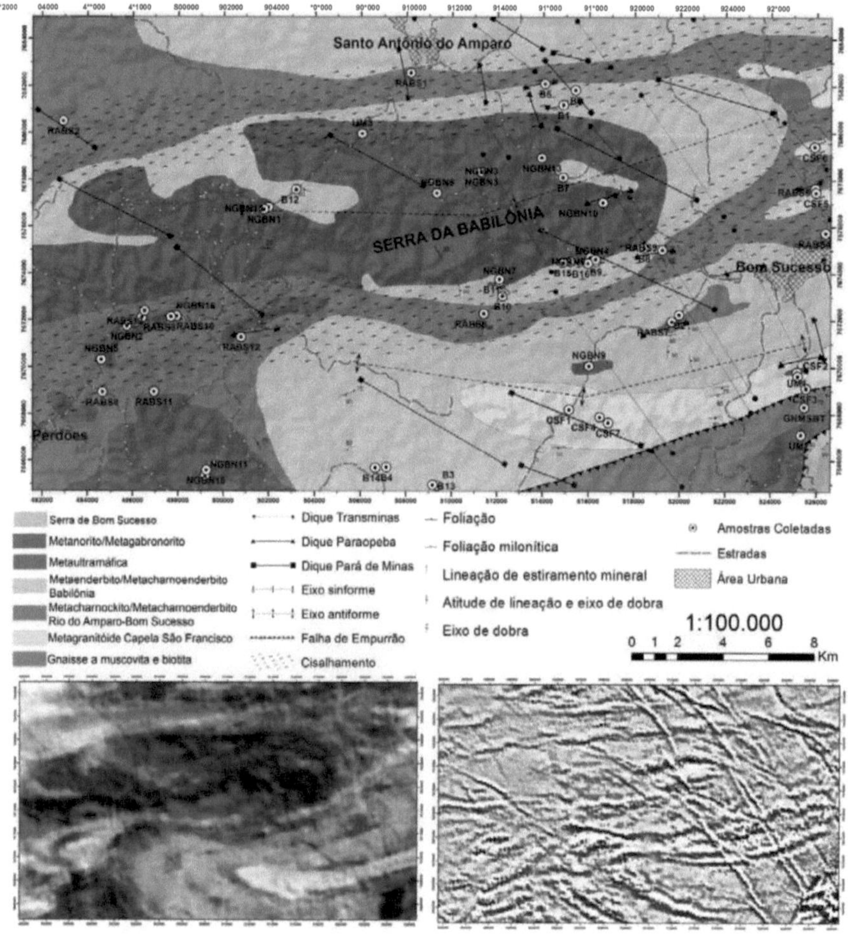

Figure 3 - Suggested geological map for the study area. Below, aerogeophysical maps, on the left the ternary false colour map (K - Th - U) and on the right the map of the 2ª vertical derivative, taken from Lasa Engenharia e Prospecções (2001).

Data from Carneiro *et al.* (2007) indicate that the local TTG gneisses have a calcium-alkali affinity and have undergone metamorphism in some portions in the amphibolite facies and in others in the granulite facies. Zircons were dated by U-Pb SHRIMP which revealed three age groups interpreted as episodes of melt generation (Teixeira *et al.,* 2000): 1- 3205 ± 17 Ma, interpreted as a reliquary age (zircon xenocrysts); 2- 3047 ± 25 Ma, interpreted as the age of the main magmatic event; 3- 2839 ±17 Ma, considered to be the age of migmatisation. Bedded ultramafic (peridotite and pyroxenite serpentinised) and mafic (metagabbro) rocks found in the area show alternation of more prominent thin sheets/layers of metapyroxenite and less prominent ones of metaperidotite (Paciullo et *al.,* 2007). Data from Carneiro et *al.* (1997) show a Sm-Nd isochron age of 2755 ± 62 Ma and Paciullo et *al.* (2007) found a Sm-Nd^M model age of 3035 Ma for these rocks. According to Pinese (1997), the metamorphised gabbro-norite bodies in the interior of the study area show an age of 2658 ± 44 Ma (Sm-Nd isochron of total rock and mineral concentrates), which he interprets as the age of intrusion. In the south-east of the study area, on the edge of the Serra de Bom Sucesso, there is the Morro das Almas Ultramafic Massif. The rocks of the massif are mainly formed by serpentine with rare traces of olivine. Its eastern portion was involved in the shear zone that generated the mountain range. As a result, the eastern half of the massif is schistified and has schist parallel to the orientation of the Serra de Bom Sucesso. Next to this is a tectonic lens of strongly millonitised TTG gneiss (Quéméneur et al., 2003).

For the metaenderbites/metachamoenderbites of the complex, called Babilónia, and represented by slightly foliated mesocratic rocks ranging from medium to coarse grained, Miranda & Chaves (2015) found an age of 2657 ± 36 Ma in monazites. This age *is* close to the 2730 Ma (Pb-Pb, zircon evaporation) found by Quéméneur et *al.* (2003) for similar local rocks.

As for the lithotypes that vary from chamockites to chamenderbites, they were called Rio do Amparo by Paciullo et *al.* (2007) and show an age close to 2718± 13 Ma (age obtained from contact relations) and Bom Sucesso by Quéméneur et *al.* (2003), who dated them at 2708 Ma (Rb-Sr). The Rio do Amparo granitoid is described by Carneiro et *al.* (2007) as a porphyritic orthogneiss with calcium-alkaline affinity, meso- to leucocratic, grey to whitish in colour, which is metamorphosed into an amphibolite-granulite facies. Depending on the intensity of the deformation, it has metamorphic/tectonic foliation that can vary from subtle to strongly millonitic. Using the Sm-Nd method, Paciullo et *al.* (2007) found a model age of 2883 Ma (age of extraction of the igneous protolith). The Bom Sucesso Granitoid has the same petrographic features as the Rio do Amparo Granitoid, is incipiently gneissified and was dated by Quéméneur et *al.* (2003) at 2708 Ma (Rb-Sr isochrone). Both describe the presence of orthopyroxene in the composition of the rock, which suggests that they are chamockitic rocks.

Quéméneur *et al.* (2003) describe a body of gneiss that occurs in a narrow, EW-orientated strip between Serra de Bom Sucesso and Fazenda São Sebastião, with a length of 15 km. The gneiss is a leucocratic rock with potassium feldspar crystals measuring 3 to 6 cm, called Capela São Francisco metagranitoid. This body has not yet been dated, but due to contact relationships it is considered to be Archaean, and is related to the potassium-rich granitic bodies from around 2700 Ma described by Romano et *al.* (2013) in the southern São Francisco Craton.

In the contact between the rocks of the Capela São Francisco metagranitoid and the ultramafic rocks described above, small muscovite nodules were observed that correspond to possible andalusite pseudomorphs (Figure 4). This suggests that the Capela São Francisco metagranitoid already existed when the ultramafic intrusion took place, leading to the growth of the nodules in the gneiss by contact metamorphism.

Figure 4 (A) Occurrence of muscovite nodules (possible andalusite pseudomorphs) at the contact between the Capela São Francisco metagranitoid and the metaultramafic. The circle shows a nodule. (B) Detail of a nodule with a prismatic habit.

MATERIALS AND METHODS

In order to determine the geological context of the area and the petrographic and lithogeochemical aspects of the rocks investigated, a bibliographic survey of the area was carried out, followed by planning and a geological survey of the study region. In the laboratories of the Institute of Geosciences of the Federal University of Minas Gerais (IGC/UFMG), 78 thin sections of rock were made for petrographic studies under a microscope, as well as pulverisation in a tungsten pan mill for lithochemical studies of the material selected from the petrography. Thirty samples of pulverised rock were sent to the SGS-Geosol Laboratory (Vespasiano-MG), where, after fusion with lithium metaborate and dilute nitric digestion, the major elements and five trace elements (Ba, Nb, Sr, Y, Zr) were analysed by ICP-OES (Inductively Coupled Plasma Optical Emission Spectrometry) and nine other trace elements, together with fourteen rare earth elements, were analysed by ICP-MS (Inductively Coupled Plasma Mass Spectrometry). The detection limits were generally around 0.01% for the larger oxides and 1 ppm for the others. The precision of the analyses is in the range 1-2% of the RSD (relative standard deviation) and the loss on ignition occurred by mass difference after heating to 1,000°C. The Geoplot programme (Zhou & Li, 2006), a Microsoft Excel add-in, was then used to analyse the geochemical data obtained.

RESULTS

PETROGRAPHY AND LITHOCHEMISTRY

From the samples collected in the field, it was possible to petrographically and lithochemically differentiate five geological units: biotite-muscovite gneiss, metaenderbito/metachamoenderbito Babilónia, metachamockito/metachamoenderbito Rio do Amparo-Bom Sucesso, metagranitoide Capela São Francisco, metaultramáfica- metanorito/metagabronorito, as well as two generations of mafic dykes.

BIOTITE-MUSCOVITE GNEISS

This unit is found in the south-east of the area. Macroscopically, it is a banded rock (Figure 5A). Microscopically, the rock has a granoblastic (Figure 5C and D) to granolepidoblastic texture, with foliation marked by felsic and mafic minerals. The most abundant mineral is quartz and its crystals are recrystallised with undulating extinction, eventually appearing in the shape of a ribbon. The contact between the grains varies from interlobed to polygonal. Microcline occurs alongside quartz, as mineral aggregates in the rock matrix and also as porphyroclasts. Most of the plagioclase is saussuritised and also appears as porphyroclast (Figure 5B). Muscovite appears together with biotite marking the foliation and some larger crystals are seen in the felsic portion, with no apparent orientation. Titanite, zircon, apatite and opaques can be seen as accessories. This unit can be classified as a biotite-muscovite gneiss.

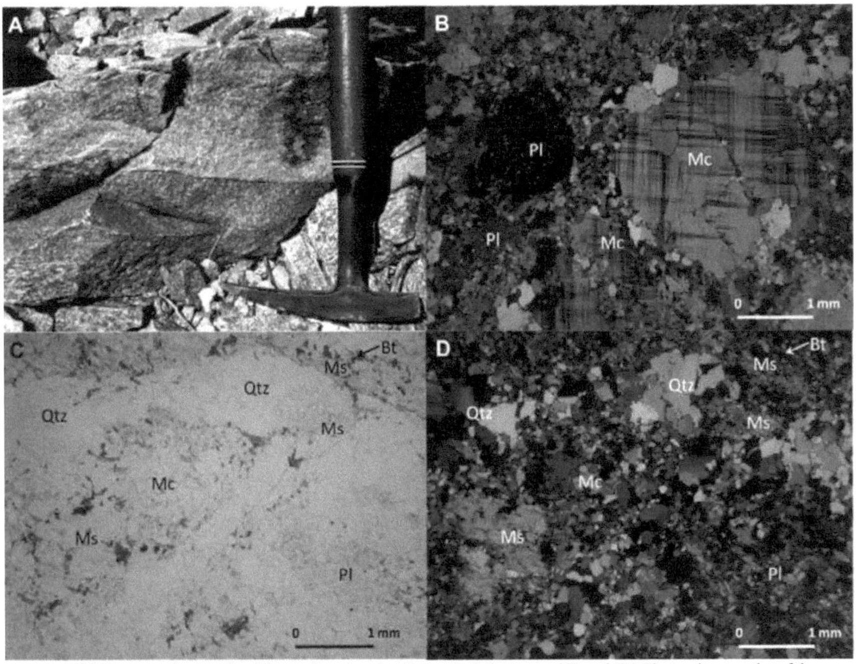

Figure 5. (A) Macroscopic sample of the gneiss, showing its banded character. (C) and (D) Photomicrographs of the general appearance of the unit under crossed polarisers (C) and crossed polarisers (D). (B) Detail of the microcline and plagioclase porphyroclasts under crossed polarisers. Qtz - Quartz, Mc - Microcline, Pl - Plagioclase, Ms - Muscovite, Bt - Biotite.

METAGRANITOID CHAPEL OF ST FRANCIS

Corresponding to the potassic arcuate intrusions described by Romano *et al.* (2013), Quéméneur et *al.* (2003) describe it as Capela São Francisco Granitic Gneiss. Macroscopically it is a mesocratic to leucocratic rock, with a granolepidoblastic texture and foliation marked by mafic minerals (Figure 6A). Microscopically, it is made up of the following minerals: quartz, microcline, plagioclase and biotite (Figure 6C and D). The accessories are garnet, cordierite, apatite, zircon and rarely muscovite. Allanite replaces garnet (Figure 6B). Quartz appears with undulating extinction and contacts varying from interlobed to polygonal. Microcline and plagioclase make up granoblastic aggregates in the rock and sometimes appear as porphyroclasts. Microcline can have a myrmecitic texture at its edge and plagioclase appears saussuritised. The biotite marks an incipient foliation and is sometimes chloritised.

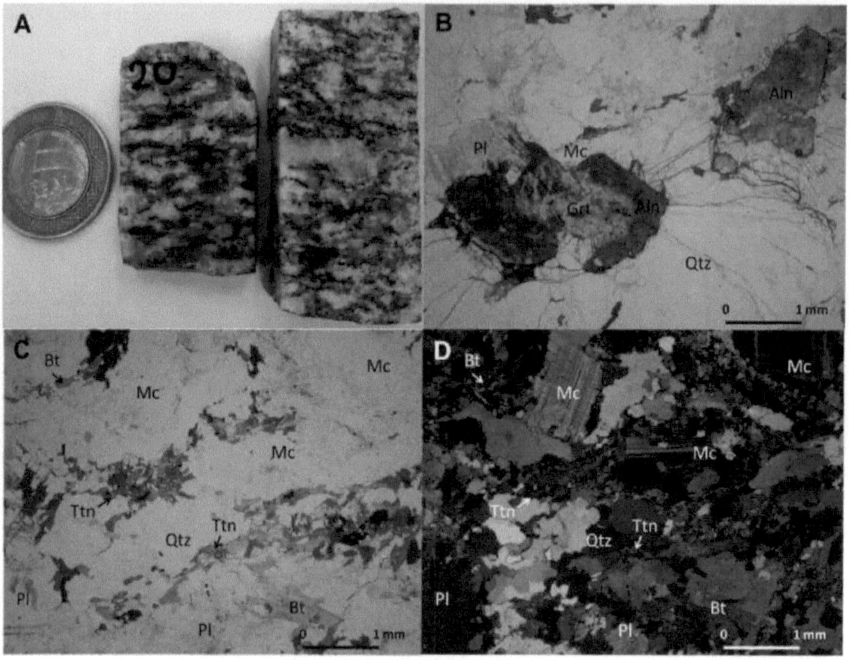

Figure 6. (A) Macroscopic sample of the unit. (C) and (D) General aspect of the unit in photomicrography with uncrossed polarisers (C) and crossed polarisers (D). (B) Detail of garnet changing to allanite in crossed polarisers. Qtz - Quartz, Mc - Microcline, Pl - Plagioclase, Bt - Biotite, Ttn - Titanite, Aln - Allanite, Grt - Garnet.

In the QAP modal diagram (Figure 7A), the Capela São Francisco metagranitoid is classified as granite.

Table 1 shows the analytical data for major elements (% by weight), trace elements and rare earths (ppm) for 6 samples of the Capela São Francisco metagranitoid. It also shows their data from the CIPW standard.

Table 1. Representative composition of the major elements, CIPW standard, trace elements and rare earths in the samples of the Capela São Francisco Metagranitoid occurring in the region studied (oxides in % by weight; trace elements and rare earths in ppm).

Metaaranitoid Cinnamon San Francisco						
	CSF1	CSF2	CSF3	CSF4	CSF5	CSF6
Major Elements						
SiO_2	76,00	74,81	77,75	75,60	75,39	74,52
TiO_2	0,21	0,23	0,26	0,28	0,34	0,27
Al2O3	11,89	12,42	11,55	12,56	13,03	12,89
FeO_T	1,51	1,38	1,74	1,76	2,04	1,61
MgO	0,22	0,23	0,26	0,30	0,25	0,32
MnO	0,02	0,01	0,02	0,02	0,01	0,01
CaO	0,96	0,97	0,62	1,01	0,33	1,22
K2O	3,96	4,19	4,44	4,56	5,31	4,70
Na_2O	3,52	3,54	3,01	3,37	1,88	3,34
P2O5	0,04	0,04	0,03	0,05	0,06	0,06
	0,01	0,01	0,01	0,01	0,01	0,01
PF	0,47	0,59	0,63	0,66	2,04	0,47
Total	98,82	98,42	100,32	100,18	100,68	99,42
CIPW standard						
Q	37.61	35.66	40.47	34.97	42.34	33.39
C	0.17	0.40	0.74	0.37	3.79	0.24
Or	23.80	25.31	26.32	27.08	31.81	28.07
Ab	30.29	30.62	25.55	28.66	16.13	28.57

An	4.57	4.65	2.89	4.71	1.26	5.72
Hy(MS)	0.56	0.59	0.65	0.75	0.63	0.81
Hy(FS)	2.50	2.22	2.81	2.82	3.25	2.56
Il	0.41	0.45	0.50	0.53	0.66	0.52
Ap	0.10	0.10	0.07	0.12	0.14	0.14
Trace elements						
Rb	283,20	216,20	275,80	265,30	173,50	195,30
Ba	289,00	394,00	172,00	444,00	482,00	750,00
Mr	67,00	93,00	40,00	89,00	52,00	129,00
Zr	102,00	118,00	148,00	162,00	174,00	183,00
Nb	25,01	18,70	26,95	20,04	4,13	9,98
Y	158,78	410,13	104,82	27,26	80,14	82,72
Ni	12,00	11,00	12,00	13,00	15,00	8,00
Co	132,10	112,40	139,10	137,10	128,60	160,30
Hf	3,87	4,46	5,14	5,19	5,17	5,40
Ta	2,72	3,23	4,23	3,20	0,88	1,25
Th	26,50	23,50	39,60	24,70	43,30	30,80
U	11,33	7,49	12,18	14,20	5,22	5,76
Rare Earth Elements						
La	69,00	475,60	54,50	26,80	82,30	95,00
Ce	44,40	56,30	67,70	59,10	91,50	119,50
Pr	12,80	84,19	11,42	5,65	13,19	16,73
Nd	48,60	297,60	41,40	19,70	49,00	56,70
Sm	10,10	58,90	8,80	4,20	8,90	9,60
Me	1,26	5,36	0,52	0,40	1,44	0,82
Gd	15,01	66,70	10,39	4,10	11,36	9,23
Tb	2,66	10,54	1,93	0,72	1,69	1,37
Dy	18,38	64,11	13,22	4,39	10,83	8,37
Ho	4,42	12,81	2,95	0,91	2,42	1,89
Er	14,53	36,14	9,40	2,84	7,46	5,70
Tm	2,15	5,12	1,41	0,44	1,08	0,74
Yb	12,60	32,20	9,20	3,00	6,60	3,60
Lu	1,95	4,48	1,37	0,45	1,03	0,51

The data (% by weight) of major elements shows that metagranitoid has between 74 and 78% $SÍO_2$, between 11 and 13% Al_2O3, FeO j less than 3%, CaO less than 1.4% and MgO less than 0.4%. They are rich in potassium (K2O between 3.5% and 5.4%) and sodium (Na2O up to 3.6%). In the An-Ab-Or normative diagram (Figure 7B) they are confirmed as granite. They are peraluminous rocks according to the ACNK-ANK diagram (Figure 7C), a feature corroborated by the presence of normative corundum in all the samples. According to the MALI (Na2O + K2O - Ca O) *versus* silica diagram (Figure 7D), they are calcium-alkaline rocks that plot in the field of peraluminous leucogranites according to Frost *et al.* (2001).

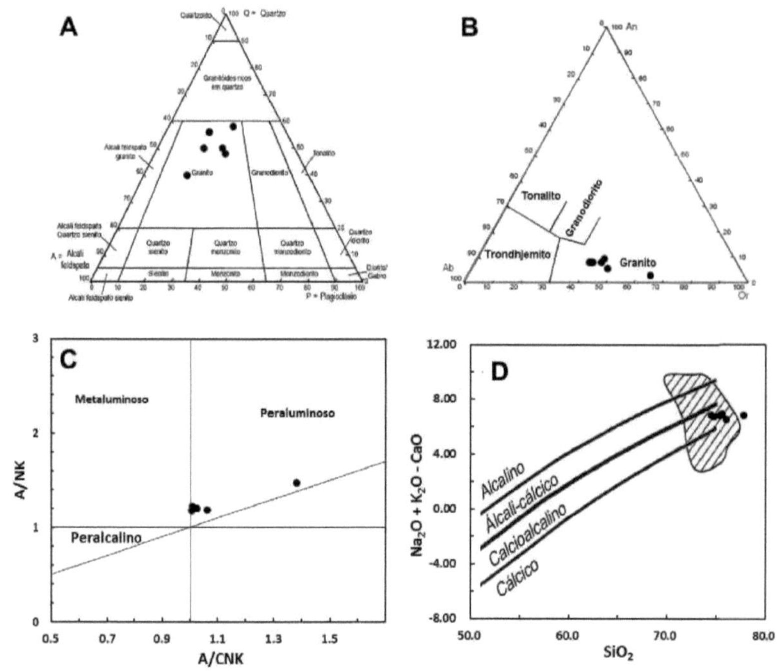

Figure 7. (A) QAP modal diagram (Streckeisen, 1974A). (B) An-Ab-Or normative diagram (Barker, 1979). (C) ACNK-ANK diagram (Maniar & Piccoli, 1989). (D) MALI x Silica diagram with the field of peraluminous leucogranites (Frost *et al.*, 2001).

The rare-earth element diagram normalised to the chondrite (Figure 8A) shows similarity in the sample patterns. The Capela São Francisco metagranitoid shows fractionation of the light rare earths (ETRL) in relation to the heavy ones (ETRP), with a negative Ce anomaly in most of the samples, probably due to the high O2 fugacity in the original magma which would have oxidised $Ce^{(3+)}$to Ce^{4+}, the latter having left the magma. There is a strong negative Eu anomaly in all the samples, possibly explained by the non-melting of the calcium plagioclase from the source rock (possibly the source was the biotite-muscovite gneiss described above) during the anatexis that generated the metagranitoid. In the diagram of incompatible elements normalised to the primitive mantle (Figure 8B), the negative Ba and Sr anomalies follow the same interpretation as the Eu anomaly. Negative Nb and Ta anomalies can be generated by the absence of these elements in the source or by the retention of ilmenite in the source. A slight positive Y anomaly is attributed to the presence of garnet in the rock. It is possible to observe similarities between the ETR and incompatible element patterns of this unit with the average of the data from the biotite granite and two-mica neoarchean granite group described by Laurent et *al.* (2014).

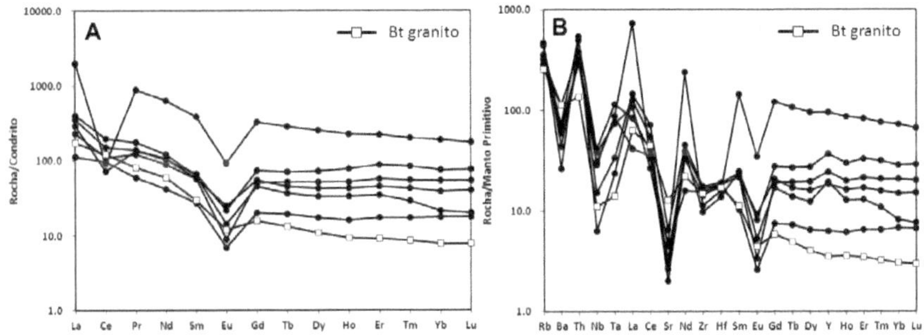

Figure 8. Arachnograms of rare earth elements (A), normalised to the chondrite according to Sun & MacDonough (1989), and incompatible elements (B), normalised to the primitive mantle according to MacDonough & Sun (1995), for the Capela São Francisco metagranitoids. The average of the biotite granite and two-mica granite data compiled by Laurent *et al.* (2014) is shown for comparison.

METACHARNOCKITO/METACHARNOENDERBITO RIO DO AMPARO-BOM SUCESSO

The petrographic and lithochemical treatment described below allowed the Bom Sucesso Granitoid described by Quéméneur et *al.* (2003) to be grouped with its similar counterpart, the Rio do Amparo Granitoid defined in the vicinity of the region studied (Carneiro et *al.*, 2007; Paciullo et *al.*, 2007). Macroscopically, it is a mesocratic to leucocratic rock with medium to coarse granulation and incipient foliation (Figure 9A). Microscopically, they are essentially made up of quartz, plagioclase, microcline and biotite, with portions of amphibole, biotite and orthopyroxene (Figure 9C and D). Titanite, zircon, opaques and rarely garnet can be seen as accessories. The presence of orthopyroxene remains classifies the rocks as chamockitoids. The quartz shows undulating extinction and the contact between the grains varies from interlobed to polygonal and appears elongated, in the shape of a "ribbon". Microcline, together with quartz, appears as granoblastic aggregates. Most of the plagioclase is saussuritised. Amphibole appears, along with a mass of quartz and opaques, replacing orthopyroxene along the foliation (Figure 9B). Biotite, when present, also marks the rock's foliation.

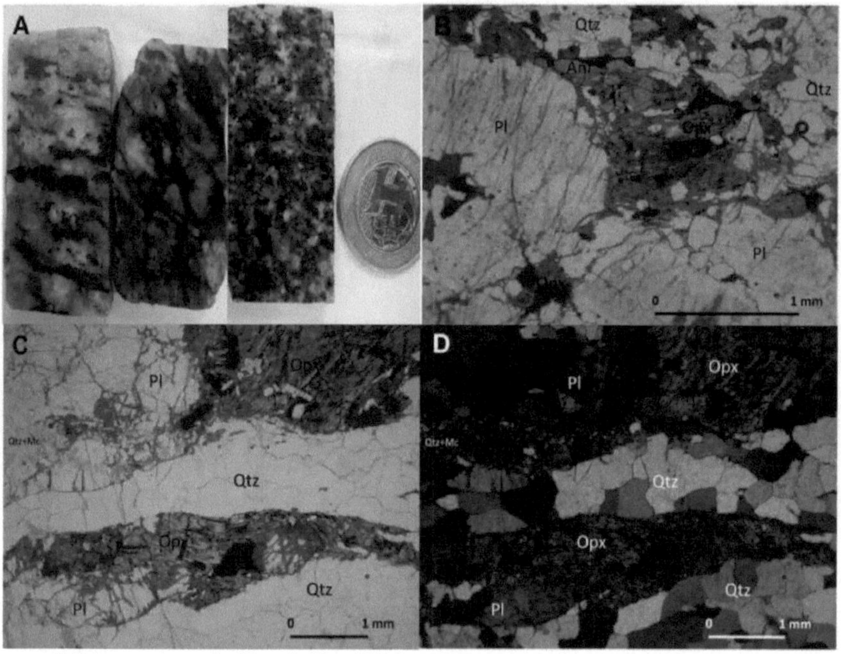

Figure 9. (A) Macroscopic sample of the unit. (C) and (D) General aspect of the unit in photomicrography with crossed polarisers (C) and crossed polarisers (D), detail of the elongated "ribbon" shaped quartz. (B) Detail of amphibole replacing orthopyroxene in crossed polarisers. Qtz - Quartz, PI - Plagioclase, Mc - Microcline, Opx - Orthopyroxene, Anf- Amphibole.

In the QAP modal diagram (Figure 10A), the unit is classified as granite and granodiorite and, as it has orthopyroxene in its composition, its lithotypes are called chamockite and chamenderbito respectively.

Table 2 shows the analytical data for major elements (% by weight), trace elements and rare earths (ppm) for 8 samples from the site. It also shows their data from the CIPW standard.

Table 2. Representative composition of the major elements, CIPW standard, trace elements and rare earths of the Metachamockito/Metachamoenderbito Rio do Amparo-Bom Sucesso samples occurring in the region studied (oxides in % by weight; trace elements and rare earths in ppm).

	Metachamockito/Metacharnoenderbito Rio do Amoaro-Bom Sucesso							
	RABS1	RABS2	RABS3	RABS4	RABS5	RABS6	RABS7	RABS8
	Major Elements							
SiO_2	70,96	72,57	72,34	70,84	68,66	66,30	72,86	71,70
TiO_2	0,69	0,40	0,50	0,66	0,93	1,39	0,76	0,41
Al2O3	12,62	12,44	12,78	12,52	12,62	12,23	12,59	12,98
FeO_T	4,18	2,61	3,28	3,09	5,05	7,10	4,52	2,93
MgO	0,46	0,21	0,33	0,64	0,83	0,95	0,64	0,55
MnO	0,03	0,01	0,03	0,01	0,05	0,06	0,01	0,02
CaO	2,26	1,20	2,09	1,71	2,30	3,07	2,38	1,38
K2O	4,11	4,64	3,56	4,82	3,32	3,55	1,83	4,17
Na_2O	3,49	3,54	3,42	3,06	3,39	3,00	4,09	3,25
P2O5	0,16	0,06	0,12	0,15	0,26	0,47	0,21	0,10
CrgOa	0,01	0,01	0,01	0,01	0,01	0,01	0,01	0,01
PF	0,34	0,50	0,56	0,54	0,63	0,16	0,88	0,66
Total	99,31	98,20	99,01	98,04	98,05	98,28	100,79	98,16
	CIPW standard							
Q	27,90	30,59	32,43	29,32	28,03	24,71	33,42	31,99
C	0,00	0,00	0,00	0,00	0,00	0,00	0,05	0,87
Or	24,54	28,07	21,37	29,22	20,14	21,38	10,83	25,28
Ab	29,84	30,67	29,40	26,55	29,45	25,87	34,65	28,21
An	6,70	4,45	9,15	6,35	9,67	9,60	10,45	6,35
Di(FS)	2,59	0,95	0,44	0,83	0,20	1,94	0,00	0,00

Di(MS)	0,52	0,14	0,08	0,33	0,06	0,48	0,00	0,00
Hy(MS)	0,92	0,47	0,80	1,48	2,09	2,19	1,60	1,41
Hy(FS)	5,28	3,74	5,10	4,28	7,93	10,03	7,07	4,86
Il	1,32	0,78	0,97	1,29	1,81	2,69	1,45	0,80
Ap	0,38	0,14	0,28	0,36	0,62	1,11	0,49	0,24
Trace elements								
Rb	76,10	125,00	85,00	146,20	110,50	67,30	80,50	245,00
Ba	1417,00	1133,00	1450,00	996,00	995,00	1191,00	473,00	1006,00
Mr	199,00	135,00	218,00	159,00	212,00	229,00	209,00	192,00
Zr	439,00	294,00	385,00	385,00	313,00	475,00	525,00	264,00
Nb	15,54	14,18	17,48	17,72	20,53	25,81	14,30	29,68
Y	34,65	33,07	29,73	30,02	48,56	58,08	29,81	38,15
Ni	14,00	12,00	20,00	17,00	17,00	10,00	14,00	16,00
Co	92,70	98,50	114,80	69,50	88,20	58,30	99,60	103,80
Hf	10,30	8,00	8,95	9,78	7,79	11,50	13,19	7,81
Ta	1,47	1,66	1,54	0,87	1,52	1,61	0,88	0,91
Th	8,40	17,50	11,90	19,90	14,30	2,80	15,40	32,50
U	0,59	7,55	1,00	1,38	2,45	1,55	1,43	2,31
Rare Earth Elements								
La	73,90	64,40	103,50	107,20	49,20	82,50	101,70	78,20
Ce	122,10	111,70	199,20	198,60	100,10	162,80	200,90	137,50
Pr	12,68	11,74	18,25	20,26	12,41	18,67	19,20	14,22
Nd	46,70	41,90	62,60	68,80	49,00	72,10	64,90	48,70
Sm	8,20	7,70	10,00	11,40	10,40	14,60	10,30	9,10
Me	1,81	1,40	1,97	0,98	1,66	2,20	1,13	1,10
Gd	7,72	7,01	8,08	9,24	9,86	13,17	8,88	9,29
Tb	1,14	1,08	1,10	1,27	1,53	2,00	1,19	1,44
Dy	6,69	6,19	6,04	6,43	9,16	11,67	6,04	7,90
Ho	1,30	1,23	1,10	1,19	1,82	2,22	1,10	1,44
Er	3,71	3,57	3,04	3,14	5,14	6,14	3,00	3,74
Tm	0,51	0,49	0,39	0,41	0,74	0,87	0,36	0,41
Yb	3,20	3,30	2,40	2,30	4,30	5,50	2,10	2,20
Lu	0,47	0,49	0,34	0,32	0,60	0,76	0,28	0,27

The data (% by weight) of major elements shows that the unit has between 66 and 74% SÍO2, between 12 and 13% AI2O3, FeOj less between 2.5% and 7.5%, CaO less than 3.1% and MgO less than 1.0%. They are rich in potassium (K2O around 4%) and sodium (Na2O between 3.0% and 4.0%). Analysing the CIPW norm data, there is minimal normative corundum in only two samples and a notable presence of normative diopside and hypersthene. In the An-Ab-Or normative diagram (Figure 10B) they are between the granite, granodiorite and trondhjemite fields. They are metaluminous rocks according to the ACNK-ANK diagram (Figure 10C), with only one sample in the peraluminous field. According to the MALI *versus* silica diagram (Figure 10D), they are mostly calcium-alkaline.

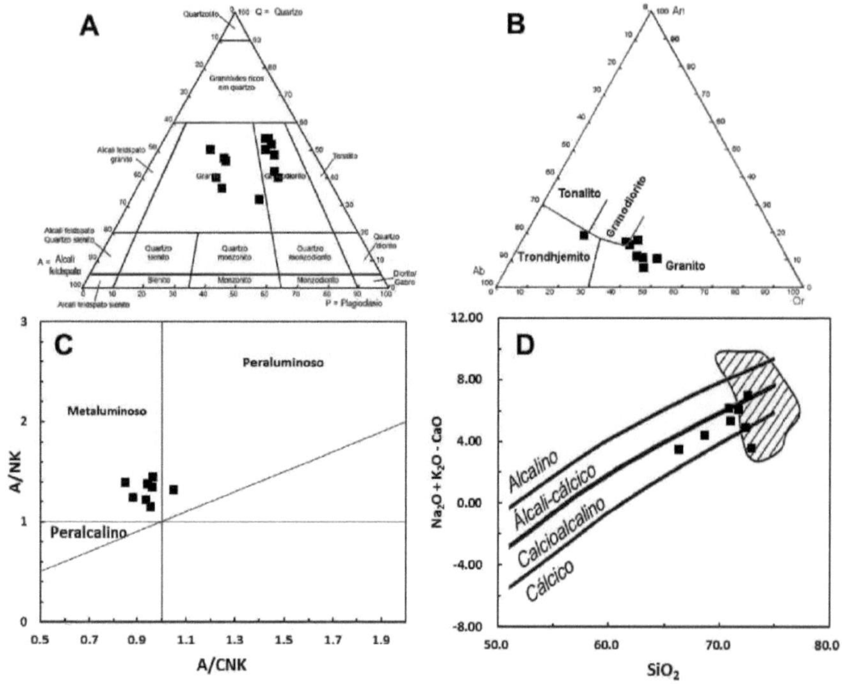

Figure 10. (A) QAP modal diagram (Streckeisen, 1974A). (B) An-Ab-Or normative diagram (Barker, 1979). (C) ACNK-ANK diagram (Mamar & Piccoli, 1989). (D) MALI x Silica diagram with the field of peraluminous leucogranites (Frost *et al.*, 2001).

The rare earth element diagram normalised to the chondrite (Figure 11 A) shows similarity in the sample patterns, with similar fractionation of ETRL and ETRP. There is a negative Eu anomaly in all the samples, possibly explained by the non-melting of the calcium plagioclase in the source rock(s). In the incompatible elements diagram (Figure 11B), the negative Sr anomaly follows the same interpretation as the Eu anomaly. Negative Nb and Ta anomalies may have been generated due to the absence of these elements in the source(s) or the retention of ilmenite in the source(s). A slight negative Y anomaly is attributed to the retention or non-existence of garnet in the source(s). It is possible to observe the similarity between the patterns of ETR and incompatible elements of this unit with the average data of the sanukite and hybrid rock types described by Laurent et *al.* (2014).

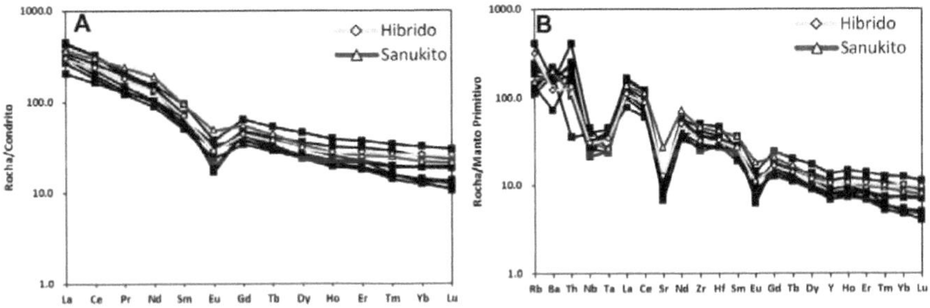

Figure 11. Arachnograms of rare earth elements (A), normalised to the chondrite according to Sun & MacDonough (1989), and incompatible elements (B), normalised to the primitive mantle according to MacDonough & Sun (1995), for the Rio do Amparo-Bom Sucesso metachamockites/metachamoenderbites. The average of the sanukite and hybrid granitoid data compiled by Laurent *et al.* (2014) is shown for comparison.

METAENDERBITO/METACHARNOENDERBITO BABILÓNIA

This unit is made up of medium to coarse-grained mesocratic rocks with a more or less accentuated mineral orientation (Figure 12A). In the southern part of the area, the rocks appear epidotised. Samples of almost isotropic rocks can be seen on slides, as well as deformed and recrystallised samples. The metaenderbites and metachamoenderbites have mineralogy composed of quartz, plagioclase, amphibole, biotite and orthopyroxene (Figure 12C and D). In the recrystallised samples, the orthopyroxenes are replaced by amphibole or biotite. Accessory minerals include monazite, zircon and opaque minerals. Microcline appears in small quantities and sometimes only as an accessory. The quartz is recrystallised, showing undulating extinction and with contacts between the grains varying from interlobed to polygonal. The plagioclase is slightly saussuritised and has a myrmecitic texture when in contact with the microclimate. In some cases, the plagioclase has a *bleb-like* texture (exsolution of potassium feldspar), probably because it remains for a long time under the action of granulite facies conditions (Mackenzie et *al.,* 1982), under which small bubbles appear inside the plagioclase (Figure 12B). The orthopyroxene appears dispersed in the slide. Amphibole and biotite are replacing orthopyroxene and when the rock is deformed they mark the foliation with a nematoblastic and lepidoblastic texture respectively. Locally, these rocks are epidote, which is why their colour is greenish.

Figure 12. (A) Macroscopic sample of the unit. (C) and (D) Photomicrograph of the general appearance of the slides from the unit with uncrossed polarisers (C) and crossed polarisers (D). (B) Photomicrograph with crossed polarisers showing *bleb-type* textures in the plagioclase and myrmequite in contact with microcline. Qtz - Quartz, PI - Plagioclase, Mc - Microcline, Opx - Orthopyroxene, Anf- Amphibole.

In the QAP modal diagram (Figure 13A), the unit is classified as granodiorite and tonalite, however, as it has orthopyroxene in its composition, they are called chamoenderbito and enderbito respectively.

Table 3 shows the analytical data for major elements (% by weight), trace elements and rare earths (ppm) for 4 samples from the site. It also shows their data from the CIPW standard.

Table 3. Representative composition of the major elements, CIPW standard, traces and rare earths (oxides in % by weight; traces and rare earths in ppm).

Metaenderbitos/Metacharnoenderbitos Babylon				
	B1	B2	B3	B4
Major Elements				
SIO2	71,03	72,74	66,77	66,80
TIO2	0,43	0,33	0,48	0,55
Al2O3	14,23	14,14	15,60	15,39
FeO_T	2,45	1,86	3,18	3,01
MgO	0,78	0,76	1,21	1,08
MnO	0,01	0,01	0,01	0,02
CaO	2,38	2,50	3,42	3,05
K2O	2,05	1,35	1,31	1,92
Na_2O	4,73	5,05	5,18	5,22
P_2O_5	0,15	0,07	0,28	0,26
Cf2O3	0,01	0,01	0,01	0,01
PF	0,59	0,51	1,41	1,06
Total	98,84	99,33	98,86	98,37
CIPW standard				
Q	28,60	30,97	21,53	20,22
C	0,27	0,00	0,12	0,00
Or	12,33	8,07	7,94	11,66
Ab	40,74	43,25	44,98	45,40
An	11,02	12,07	15,54	13,25

Di(FS)	0,00	0,01	0,00	0,28
Di(MS)	0,00	0,01	0,00	0,19
Hy(MS)	1,98	1,91	3,09	2,68
Hy(FS)	3,88	2,92	5,20	4,64
Il	0,83	0,63	0,94	1,07
Ap	0,35	0,16	0,67	0,62
Trace elements				
Rb	48,40	21,40	22,10	19,90
Ba	1134,00	464,00	908,00	1523,00
Mr	452,00	473,00	1305,00	1479,00
Zr	205,00	137,00	240,00	295,00
Nb	4,42	5,23	4,27	5,12
Y	7,59	4,69	13,25	24,71
Ni	14,00	17,00	17,00	13,00
Co	96,00	133,30	50,10	67,00
Hf	5,20	3,82	5,20	6,62
Ta	0,58	0,82	0,30	0,48
Th	10,70	4,30	22,10	20,30
U	0,44	0,31	0,42	0,36
Rare Earth Elements				
La	90,90	25,10	183,50	163,60
Ce	132,80	39,00	317,10	293,40
Pr	15,03	4,00	31,11	29,87
Nd	48,50	14,10	103,40	104,30
Sm	5,80	2,20	12,90	14,70
Me	1,30	0,70	2,38	3,17
Gd	3,75	1,81	7,89	10,15
Tb	0,40	0,23	0,82	1,14
Dy	1,71	1,06	3,19	5,33
Ho	0,30	0,19	0,51	0,89
Er	0,73	0,46	1,20	2,25
Tm	0,09	0,06	0,12	0,27
Yb	0,50	0,30	0,70	1,40
Lu	0,07	0,05	0,09	0,18

The data (% by weight) of major elements shows that the unit has between 66.5 and 73% SÍO2, between 14 and 16% AI2O3, minor FCOT between 1.5% and 3.5%, CaO less than 3.5% and MgO less than 1.3%. They are not rich in potassium (K2O around 2%) and have high sodium (Na20 around 5.0%). From a normative point of view, there is no or minimal presence of normative diopside and a high amount of hypersthene. In the An-Ab-Or normative diagram (Figure 13B) they are classified as trondhjemite. They are metalluminous rocks according to the ACNK-ANK diagram (Figure 13C). According to the MALI *versus* silica diagram (Figure 13D), they range from calcic to calc-alkaline.

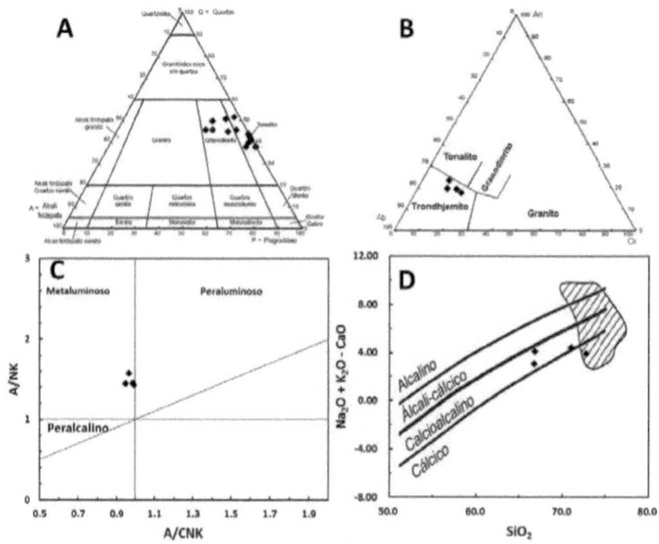

Figure 13. (A) QAP modal diagram (Streckeisen, 1974A). (B) An-Ab-Or normative diagram (Barker, 1979). (C) ACNK-ANK diagram (Mamar & Piccoli, 1989). (D) MALI x Silica diagram, with the field of peraluminous leucogranites (Frost *et al.*, 2001).

The rare earth element diagram normalised to the chondrite (Figure 14A) shows similarity in the patterns of the samples. The Babylon meta-enderite/metachamockite shows a similar fractionation of ETRLs to the ETRPs, with no marked anomalies. The diagram of incompatible elements (Figure 14B) shows a marked anomaly of Nb and Ta. It is possible to observe similarities between the ETR and incompatible element patterns of this unit with the average of the TTG-type data described by Laurent et *al.* (2014).

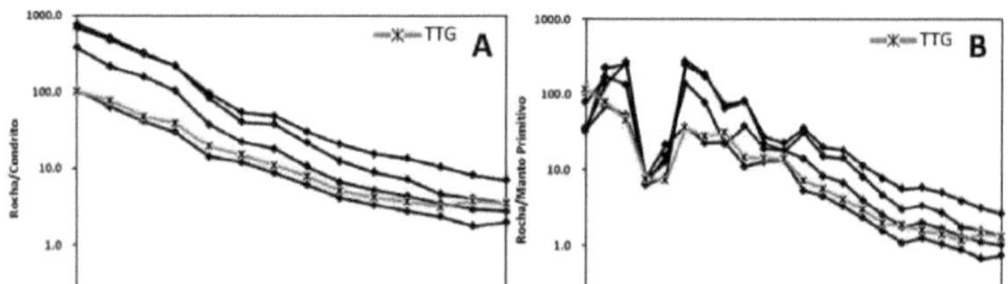

Figure 14. Arachnograms of rare earth elements (A), normalised to the chondrite according to Sun & MacDonough (1989), and incompatible elements (B), normalised to the primitive mantle according to MacDonough & Sun (1995), for the Babylon metaenderbites/metachamoenderbites. The average TTG data compiled by Laurent et *al.* (2014) is shown for comparison.

METAULTRAMAFIC

This unit occurs in small portions of the study area, with an apparent correlation with the bedded mafic-ultramafic Ribeirão dos Motas Sequence defined by Carneiro *et al.* (2007), which appears in the vicinity of the studied area. It shows a bedded structure and cumulatic texture, made up of pyroxene bands and serpentinised olivine bands. Spinel and opaque minerals can be seen as accessories (Figure 15). It can be classified as a metaperidotite.

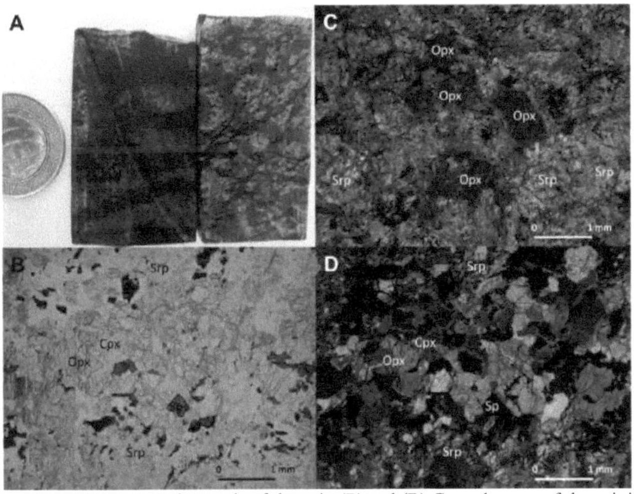

Figure 15. (A) Macroscopic sample of the unit. (B) and (D) General aspect of the unit in photomicrographs with uncrossed polarisers (B) and crossed polarisers (D), with serpentinised olivine cumulus and pyroxene cumulus with spinel. (C) Detail of serpentinised olivine pseudomorphs. Opx - Orthopyroxene, Cpx - Clinopyroxene, Srp - Serpentine, Sp - Spinel.

METANORITE/METAGABRONORITE

The lithotypes of this unit were described by Quéméneur et *al.* (2003) in the Serra da Babilónia region (Figure 16A) and occur in association with ultramafic rocks. They generally appear with a preserved igneous texture (Figure 16C and D), but recrystallised portions are also observed (Figure 16B). They have the following mineralogy: plagioclase, orthopyroxene and clinopyroxene in varying relative proportions, and small amounts of amphibole, biotite and quartz. Opaque minerals are found as accessories. Because the proportions of pyroxenes vary, these rocks can correspond to compositions between gabbro and norite. Plagioclase is the majority in the unit, while amphibole and biotite appear at the edge of the pyroxenes.

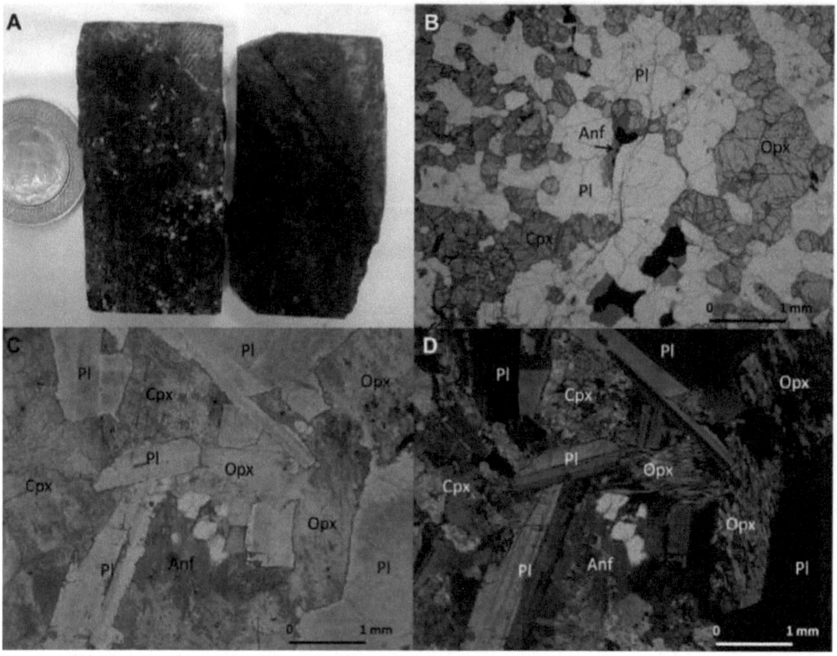

Figure 16: (A) Macroscopic sample of the unit. (C) and (D) Photomicrographs of the rock with preserved igneous texture with uncrossed polarisers (C) and crossed polarisers (D). (B) Photomicrograph with crossed polarisers showing the rock when recrystallised. Opx - Orthopyroxene, Cpx - Clinopyroxene, PI - Plagioclase, Anf - Amphibole.

Chemically, 12 samples were analysed, 1 from ultramafic rock and 11 from metamorphic and metagabronoritic (mafic) rocks (Table 4). They show no petrographic evidence of hydrothermal or weathering alteration. The mafic rock samples generally showed fire loss values of less than 1%, while the ultramafic rock sample reached a value of 10% due to the high degree of serpentinisation.

Table 4. Representative composition of major elements, CIPW standard, trace elements and rare earths in samples of occurring in the region studied (oxides in % by weight; traces and rare earths in ppm).

	Metaultramafic	Metanorites/Metaaabronorites										
	UM1	NGBN1	NGBN2I	NGBN3	NGBN4	NGBN5	NGBN6	NGBN7I	NGBN8	NGBN9	NGBN10	NGBN11
	Major Elements											
SlO2	44,15	55,36	53,63	51,24	52,13	55,71	53,05	50,61	50,55	49,16	51,32	53,76
TIO2	0,16	0,63	1,02	1,56	1,26	0,38	0,48	0,87	0,48	1,90	1,03	0,39
Al2O3	4,37	11,09	11,77	13,18	12,76	15,64	11,11	14,48	11,40	13,53	13,19	11,55
FeO_T	6,99	9,88	13,57	13,68	13,47	8,51	11,21	13,17	11,91	16,87	13,63	8,90
MgO	30,85	11,02	6,86	6,09	6,45	5,65	13,91	7,07	13,67	4,67	6,53	13,89
MnO	0,12	0,15	0,21	0,20	0,21	0,25	0,21	0,24	0,21	0,27	0,21	0,14
CaO	2,41	8,20	8,10	9,78	10,10	10,02	8,30	10,48	8,20	9,21	9,52	7,93
K2O	0,02	0,54	0,24	0,25	0,25	0,62	0,42	0,51	0,48	0,38	0,44	0,55
Na_2O	0,01	1,61	2,83	2,23	2,17	1,24	1,46	2,20	1,54	2,38	1,97	1,52
P_2O_5	0,01	0,05	0,09	0,15	0,08	0,11	0,10	0,13	0,09	0,22	0,17	0,05
O2O3	0,33	0,15	0,05	0,03	0,02	0,21	0,20	0,00	0,19	0,00	0,01	0,27
PF	9,37	0,54	0,14	0,01	0,01	1,29	0,24	0,43	0,26	0,01	0,43	0,51
Total	98,79	99,22	98,51	98,40	98,91	99,63	100,69	100,19	98,98	98,60	98,45	99,46
Mg#	91,1	72,2	54,1	50,9	52,7	60,7	74,3	55,6	72,8	39,2	52,8	78,4
	Noma CIPW											
Q	0,00	6,34	2,72	2,12	2,50	12,69	0,00	0,00	0,00	0,00	2,14	1,45
Or	0,13	3,24	1,44	1,50	1,50	3,74	2,48	3,02	2,88	2,28	2,65	3,29
Ab	0,09	13,83	24,35	19,18	18,57	10,70	12,32	18,66	13,23	20,43	17,01	13,03
An	13,27	21,76	19,02	25,64	24,61	35,95	22,47	28,20	23,12	25,47	26,37	23,38
Di(FS)	0,01	5,57	9,71	10,82	11,88	5,68	4,92	10,30	5,08	11,25	9,70	3,74

Di(MS)	0,05	10,08	8,06	8,22	9,50	5,93	9,68	8,97	9,25	5,30	7,63	9,31
Hy(MS)	43,21	23,18	13,64	11,61	11,84	11,60	28,71	9,55	21,15	8,17	13,06	30,75
Hy(FS)	7,20	14,68	18,87	17,54	16,99	12,74	16,72	12,58	13,34	19,91	19,04	14,18
Ol(MS)	30,14	0,00	0,00	0,00	0,00	0,00	0,96	2,77	6,39	0,82	0,00	0,00
Ol(FS)	5,53	0,00	0,00	0,00	0,00	0,00	0,61	4,02	4,44	2,21	0,00	0,00
Il	0,34	1,21	1,97	3,01	2,42	0,74	0,91	1,66	0,93	3,66	2,00	0,75
Ap	0,03	0,12	0,21	0,35	0,19	0,26	0,23	0,30	0,21	0,52	0,40	0,12
					Trace elements							
Rb	0,70	15,00	1,60	4,30	6,30	19,10	8,60	16,20	13,90	7,00	14,90	24,00
Ba	36,00	155,00	116,00	63,00	59,00	104,00	129,00	129,00	135,00	105,00	116,00	174,00
					103,0							
Mr	10,00	108,00	134,00	104,00 0		140,00	88,00	118,00	95,00	118,00	129,00	101,00
Zr	10,00	50,00	60,00	112,00	59,00	41,00	62,00	83,00	68,00	154,00	137,00	20,00
Nb	4,94	3,11	4,72	7,06	4,94	0,06	0,48	2,17	2,06	2,95	1,80	5,95
Y	3,82	13,44	21,88	32,24	24,78	14,16	12,18	24,38	13,89	27,08	23,66	11,19
Ni	1688,00	239,00	149,00	104,00	92,00	220,00	280,00	129,00	380,00	40,00	95,00	420,00
Co	81,30	70,60	101,40	79,80	83,00	72,30	68,30	76,00	87,30	64,10	72,00	81,20
Hf	0,05	1,60	1,87	3,17	2,05	0,70	1,12	2,02	1,49	2,27	1,95	1,08
Ta	0,32	0,31	0,64	0,67	0,66	0,08	0,08	1,25	0,48	0,25	0,26	0,34
Th	0,40	1,80	0,50	2,60	1,40	0,40	0,90	2,20	1,60	1,00	2,00	4,10
U	0,07	0,40	0,10	0,65	0,41	0,08	0,24	0,79	0,48	0,39	0,59	1,12
					Rare Earth Elements							
La	5,60	11,90	11,20	16,90	12,70	10,90	8,10	22,80	9,90	9,20	12,10	19,20
Ce	2,30	17,30	19,60	27,40	17,20	10,60	12,70	27,20	16,90	19,60	19,40	25,00
Pr	0,42	2,03	2,66	3,48	2,23	1,32	1,58	3,18	1,94	2,75	2,76	2,83
Nd	1,40	8,40	12,60	16,30	10,80	5,70	6,90	13,50	8,40	13,90	12,40	10,70
Sm	0,30	2,10	3,30	4,50	3,00	1,50	1,80	3,20	1,90	3,70	3,40	2,10
Me	0,13	0,63	1,14	1,31	1,06	0,45	0,48	1,03	0,65	1,14	1,12	0,56
Gd	0,45	2,39	4,21	5,81	3,94	2,06	1,98	3,89	2,15	4,68	3,68	2,10
Tb	0,08	0,40	0,66	0,96	0,70	0,30	0,28	0,64	0,36	0,76	0,64	0,32
Dy	0,54	2,54	4,09	6,10	4,46	2,31	2,10	4,40	2,55	5,00	4,28	1,96
Ho	0,13	0,53	0,86	1,28	0,94	0,45	0,40	0,82	0,42	0,95	0,90	0,43
Er	0,39	1,44	2,40	3,69	2,88	1,61	1,26	2,56	1,71	3,12	2,81	1,26
Tm	0,05	0,22	0,36	0,54	0,42	0,19	0,14	0,32	0,15	0,35	0,42	0,18
Yb	0,40	1,50	2,10	3,50	2,70	1,60	1,20	2,70	1,50	3,00	2,60	1,30
Lu	0,06	0,21	0,32	0,53	0,42	0,08	0,05	0,19	0,05	0,31	0,24	0,19

The arrangement of the samples on the TAS classification diagram (Le Bas *et al.,* 1986) indicates the subalkaline and silica-saturated character, where the mafic rocks are positioned in the field of basalts and andesitic basalts and the metaultramafic rock is positioned in the field of picrites (Figure 17A).

From a normative point of view, we can highlight the high presence of normative olivine in the ultramafic rock sample. The norites and gabronorites vary from quartz-tourmaline to normative olivine-tourmaline.

The arrangement of the samples on the Jensen (1976) diagram classifies the mafic metamorphic/metagabronoritic rocks as basaltic to toleitic high-Fe komatiite and the ultramafic as picritic komatiite (Figure 17B). The mafic and ultramafic rocks show a toleitic character and are positioned above the boundary of the field dividing the toleitic and calc-alkaline rocks of Kuno (1968) in the Irvine & Baragar (1971) diagram (Figure 17C). In the SiCh vs. K2O diagram (Rickwood, 1989) all the samples show low K, with the ultramafic sample having the lowest SiCh/K2O ratio. The alignment of the samples in the diagrams in Figures 17B, C and D suggests that the mafics are the product of the magmatic differentiation of the ultramafics promoted by the igneous processes that generate the cumulitic texture in the latter.

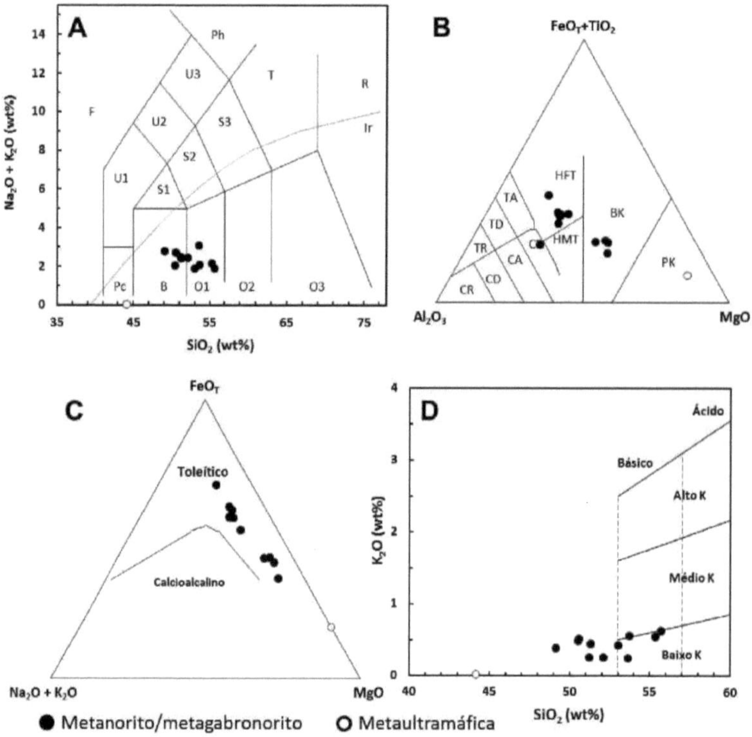

Figure 17 (A) Total alkali versus silica diagram (TAS - Le Bas *et al.*, 1986). Pc = picrites, B = basalts, O1 = andesitic basalts. (B) FeOt+TiO_2-Al_2O3-MgO diagram from Jensen (1986). PK = picritic komatiite, BK = basaltic komatiite, HFT = high Fe toleite. (C) Diagram from Irvine & Baragar (1971). (D) Diagram of SiO_2 vs. K_2O Rickwood (1989).

There is a similarity between the ETR patterns of ultramafic and mafic rocks, all of which show a slight fractionation of ETRL in relation to ETRP (Figure 18A), with a slight negative Eu anomaly in mafic rocks. This anomaly can be explained by a possible high O2 fugacity that promotes the oxidation of Eu^{2+} to Eu^{3+} in the mafic magma, with the latter escaping from the magma, leaving a negative Eu signature in it. The incompatible elements diagram (Figure 18B) shows negative Sr and Hf anomalies in the metaultramafic and mafic rocks, explained by the absence of plagioclase and zircon, respectively. The negative Nb anomalies in the metanorites/metagabronorites can be attributed to the retention of ilmenite by the ultramafic magma, which apparently differentiated itself from the mafic magma. It should be noted that the ETRP of the metamorphic is positioned below the reference line of the primitive mantle of Sun & MacDonough (1989).

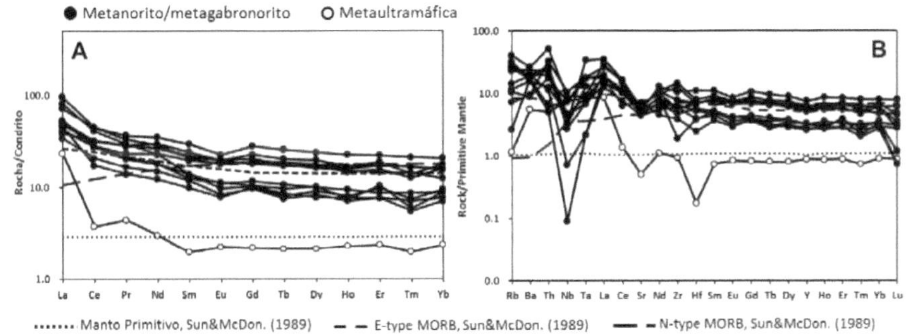

Figure 18. Arachnograms of rare earth elements (A), normalised to the chondrite according to Sun & MacDonough (1989). And of incompatible elements (B), normalised to the primitive mantle according to MacDonough & Sun (1995).

SWARMS OF DYKES

During the fieldwork, two different generations of dykes were sampled out of the three that occur within the studied area. Dykes with an amphibolitic texture were found, sometimes with plagioclase porphyroclasts, corresponding to the Paraopeba swarm of Chaves (2013). The dykes in this cluster form conjugate pairs in the N70E and N20W directions. Dykes corresponding to the Pará de Minas swarm by Chaves (2013), heading N60W, also outcrop in the area. Preserved outcrops of dykes from the Transminas swarm of N20-30W direction and Cretaceous age (Chaves, 2013) were not found in the field, even though they can be seen in the magnetometric image in figure 3.

On a slide, the samples from the Paraopeba swarm have a granonematoblastic texture and are quite recrystallised. Petrographically, they are made up of plagioclase and clinopyroxene, the latter largely replaced by amphibole during amphibolite facies metamorphism along shear zones into whose transtensional portions the magma from these dykes would have intruded. Quartz and opaques appear as accessories. Biotite and chlorite are secondary minerals formed from amphibole during regressive metamorphism (Figure 19).

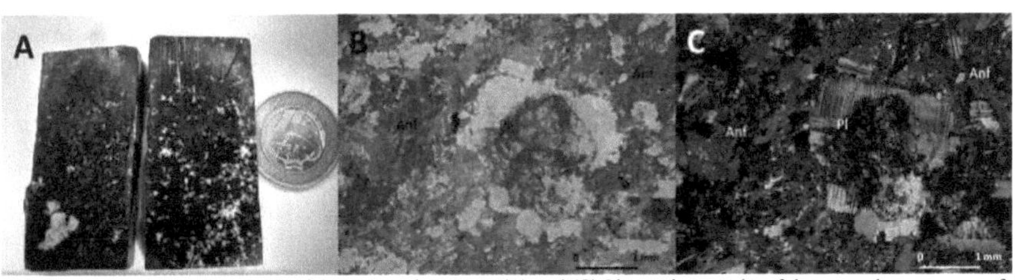

Figure 19 (A) Macroscopic sample of the Paraopeba swarm (B) and (C) Photomicrographs of the general appearance of the dykes on slides with uncrossed polarisers (B) and with crossed polarisers (C). Anf = amphibole, Pl = plagioclase.

The samples from the Pará de Minas swarm on slides are mainly composed of clinopyroxene and plagioclase. Quartz appears in small quantities and opaque minerals as accessories (Figure 20).

Figure 20. (A) Macroscopic sample from the Pará de Minas swarm. (B) and (C) Photomicrograph showing the general appearance of the dykes on slides with uncrossed polarisers (B) and with crossed polarisers (C). Cpx = clinopyroxene, PI = plagioclase.

DISCUSSION AND CONCLUSIONS

The biotite-muscovite gneiss found in the southeast of the area is apparently related to the oldest rocks in the region, known as the TTG gneisses of the Campo Belo complex, which were dated by Carneiro *et al.* (2007) to around 3000 Ma. In the study area, this gneiss seems to have been brought to the surface in the form of a tectonic splinter as described by Quéméneur et *al.* (2003) in the Serra de Bom Sucesso region.

The Capela São Francisco metagranitoid, which has not yet been dated in the study area, is related to the K-rich granitoids of crustal origin, widely found in the southern portion of the São Francisco Craton with an age of around 2750 *and* 2700 Ma, according to Romano et *al.* (2013). Chemically, it resembles the biotite granite group of Laurent et *al.* (2014) shown in figure 21. It is peraluminous and has a high silica and K content, low Ca, with ETR and incompatible element patterns similar to this group described by Laurent et *al.* (2014). These authors

attribute the origin of this lithotype to the anatexis (partial fusion) of older paraderived crust which, in the case of this study, would be the crust represented by muscovite and biotite gneisses (strong contribution of sedimentary protolith - Figure 21B). The presence, albeit rare, of cordierite in the Capela São Francisco metagranitoid suggests that this anatexis occurred under early granulite facies conditions (Yardley, 1994) during the migmatisation of the paragneissic protolith to two micas.

The Rio do Amparo-Bom Sucesso metachamockite/metachamoenderbito has chemical characteristics similar to the sanukite and hybrid granitoid groups of Laurent *et al.* (2014) (Figure 21 A), they are metalluminous and calcium-alkaline. The RTS and incompatible element patterns are similar to those of the Laurent et *al.* (2014) groups, as shown in Figure 11. The high concentrations of elements from the crust and mantle is a typical characteristic of sanukites and hybrid granitoids, showing that their origin comes from the interaction of different sources (mantle and crustal) in the lower crust (Figure 21B). This is perhaps why Quéméneur et *al.* (2003) suggested that magma mixing processes were recorded in the region's rocks.

The Babylonian metaenderite/metachamoenderite is metalluminous, rich in Na, has chemical characteristics similar to the TTG group of Laurent et *al.* (2014) (Figure 21A) and ranges from calcium to calcium-alkaline. They have a typical fractionated ETR pattern, with no Eu anomalies. Their origin is linked to the partial melting of K-poor mafic-ultramafic mantle sources (Figure 21B). The presence of orthopyroxene substituted for amphibole in the Rio do Amparo-Bom Sucesso and Babilónia units suggests that the crystallisation/recrystallisation of these rocks took place under a granulite facies, with retrometamorphism into an amphibolite facies.

The alignment of the samples in the diagrams in figure 17 suggests that the mafic metanorites/metagabronorites are the product of the magmatic differentiation of the metaultramafics promoted by the igneous processes that generate the cumulic texture in the latter. This suggestion is corroborated by the similarities in the patterns of ETR and incompatible elements in figure 18 and by the same spatial arrangement in the study area of the metaultramafic and mafic units.

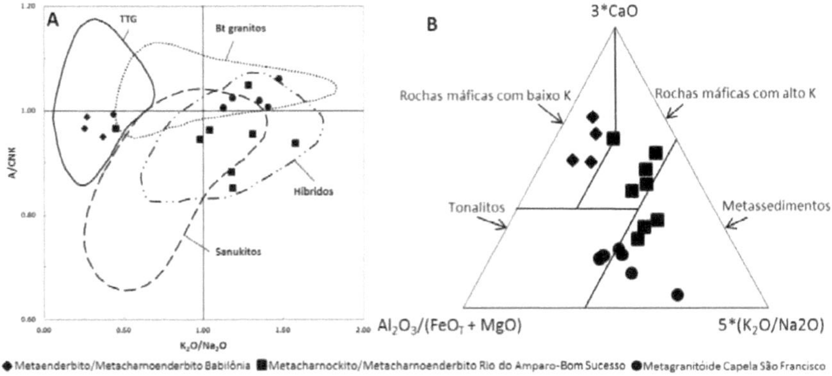

Figure 21 (A) K_2O/Na_2O x A/CNK diagram. The delimited fields represent the compilation of data from neoarchaeal granitoid rocks from various cratons around the world. (B) Ternary diagram $Al_2O_3/(FeO_T + MgO)$; 3*CaO; 5*($K_{(2)}O/Na_{(2)}O$), where the fields separated by straight lines represent the composition of melts derived from potential sources (diagrams modified from Laurent *et al.*, 2014).

It is possible to fit the rocks in the study area into the global evolutionary model for late Archean granitoids proposed by Laurent et *al.* (2014) (Figure 22). During the first phase (magmatic stage 1), there is subduction of the oceanic crust (Figure 22A) forming island arcs or continental magmatic arcs and leading to the generation of several pulses of TTG-type juvenile magmatism over a long period and the metasomatism of the mantle wedge by fluids derived from the basalts and subducted sediments. The first phase leads to the closure of an oceanic basin and the collage of two proto-continental blocks. In the next phase (magmatic stage 2), sanukitoids are formed due to significant interaction between the latest TTG fusions and peridotites, while crustal thickening initiates partial fusion of the sanukitoids formed, giving rise to biotite granites and granites with two micas (Figure 22B). Sanukitoids can also be formed during the syn- to late-collisional period due to thermal and crustal relaxation (end of magmatic stage 2), which includes the melting of the enriched lithospheric mantle (Figure 22C). During the syn- and late-collisional periods (Figure 22B and C) the interaction between the sanukitoids and the melts derived from the crust generate various types of granitoid mixtures (hybrids).

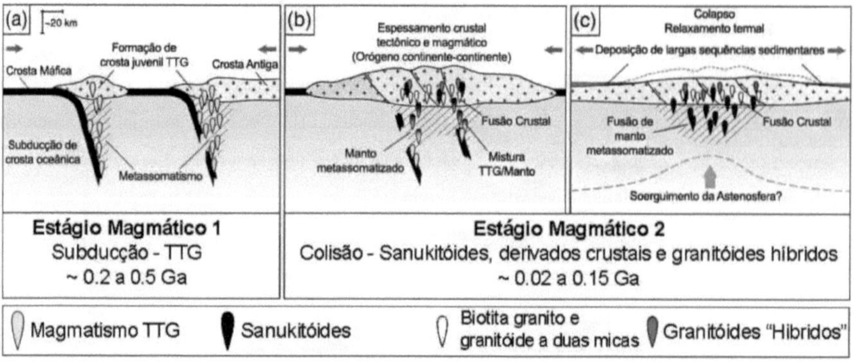

Figure 22. Conceptual model for the geodynamic evolution of late Archaean granitoids (Extracted from Laurent *et al.*, 2014).

Magmatic stage 2 of the model in figure 22 seems to be able to explain the geological processes and products of the final phase of the Rio das Velhas Neoarchean event in the southern region of the São Francisco Craton, within which the rocks of the area studied around Serra da Babilónia (Bom Sucesso-MG) chronologically fit. This stage 2 would include the five geological units described here, namely: (I) the muscovite and biotite gneiss from around 3000 Ma, whose partial fusion (anatexis) around 2700 Ma would give rise to the (II) Capela São Francisco metagranitoid (the one corresponding to the biotite granite in figure 22); the Rio do Amparo-Bom Sucesso metachamockites and metachamoenderbites from around 2700 Ma (III) would correspond to the sanukites and hybrid granitoids in figure 22; the Babilonia metaenderbites and metachamoenderbites (IV), also around 2700 Ma, would be the TTGs enriched by the metasomatised mantle; the metaultramafics as a product of partial fusion of the metasomatised lithospheric mantle (enriched in light ETR) would differentiate into the metanorites and metagabronorites (V) from around 2700 Ma. These Neoarchean geological processes and products would have occurred in the lower crust, in granulite facies conditions, close to the crust-mantle boundary equivalent to the region between the Conrad and Mohorovicic discontinuities.

The geological processes responsible for the emplacement of the late mafic dykes have no connection with those presented in the previous paragraph, having taken place at another geological moment, already in the Palaeoproterozoic and Cretaceous.

ACKNOWLEDGEMENTS

To CNPq for the scholarship awarded to the authors. To the UFMG Geosciences Institute, which made the CPMTC laboratories available to support this research.

BIBLIOGRAPHICAL REFERENCES

1. ALKMIM, F.F. What makes a craton a craton? The São Francisco Craton and the Almeidian revelations in delimiting it. In: MANTESSO-NETO, V.; BARTORELLI, A.; DAL RÉ CARNEIRO C.; BRITO-NEVES, B.B. (Org.). Geology of the South American Continent. Evolution of the work of Fernando Flávio Marques de Almeida. Becca, São Paulo, p. 17-35. 2004.

2. ALMEIDA, J.A.C.; DALLAGNOL, R.; OLIVEIRA, M.A.; MACAMBIRA, M.J.B.; PIMENTEL, M.M.; RÃMÕ, O.T.; GUIMARÃES, F.V.; LEITE, A.A.S. Zircon geochronology, geochemistry and origin of the TTG suites of the Rio Maria granite-greenstone terrane: implications for the growth of the Archean crust of the Carajás Province, Brazil. Precambrian Research. 187. p. 201-221. 2011.

3. ALMEIDA, J.A.C.; DALL'AGNOL, R.; LEITE, A.A.S. Geochemistry and zircon geochronology of the Archean granite suites of the Rio Maria granite-greenstone terrane, Carajás Province, Brazil. Journal of South American Earth Sciences. 42. p. 103-126. 2013.

4. ÁVILA, C.A.; TEIXEIRA, W.; CORDANI, U.G.; MOURA, C.A.V.; PEREIRA, R.M. Rhyacian (2.23-2.20 Ga) juvenile accretion in the Southern São Francisco Craton, Brazil: Geochemical and isotopic evidence from the Serrinha Magmatic Suite, Mineiro Belt. Journal of South American Earth Sciences. 29. p. 464-482. 2010.

5. BARKER, F. Trondhjemites: definitions, environment and hypotheses of origin. In: F Barker (eds.) Trondhjemites, dacites and related rocks. Amsterdam, Elsevier, p. 1-12. 1979.

6. CAMPOS, J.C.S.; CARNEIRO, M.A.; BASEI, M.A.S. U-Pb evidence for Late Neoarchean crustal reworking in the Southern São Francisco Craton (Minas Gerais, Brazil). Annals of the Brazilian Academy of Sciences. 75(4). p. 497-511. 2003.

7. CARNEIRO, M.A. The Bonfim Setentrional Metamorphic Complex (Iron Quadrangle, Minas Gerais): Lithostratigraphy and geological evolution of an Archean continental crust segment. PhD Thesis, University of São Paulo, São Paulo. 233 p. 1992.

8. CARNEIRO, M.A. *et al.* Sheets Campo Belo (SF.23-V-B-VI) and Oliveira (SF.23-X-A-IV), scale 1:100.000: final report. Ouro Preto. Geology of Brazil Programme: Geological Maps. 192 p. 2007.

9. CARNEIRO, M.A. & BARBOSA, M.S.C. Geological and tectonic implications of the magnetometric interpretation of the Oliveira region, Minas Gerais. Revista Brasileira de Geofísica. 26(1). p. 87-98. 2008.

10. CHAVES, A.O. Mafic dyke swarms of Minas Gerais - the state of the art. Geonomos. 21(1). p. 29-33.2013.

11. DEY, S.; NANDY, J.; CHOUDHARY, A.K.; LIU, Y.; ZONG, K. Origin and evolution of granitoids associated with the Kadiri greenstone belt, eastem Dharwar craton: a history of orogenic to anorogenic magmatism. Precambrian Research. 246. p. 64-90. 2014.

12. ENGLER, A.; KOLLER, F.; MEISEL, T.; QUÉMÉNEUR, J. Evolution of the archean/proterozoic crust in the Southern São Francisco craton near Perdões, Minas Gerais, Brazil: petrological and geochemical constraints. Journal of South American Earth Sciences 15. p. 709-723. 2002.

13. FERNANDES, R.A. & CARNEIRO, M.A. The Campo Belo Metamorphic Complex (Southern São Francisco Craton): lithodemic units and tectonic evolution. Rev. Brasileira de Geociências. 30(4). p. 671-678. 2000.

14. FROST, B.R.; BARNES, C.G.; COLLNS, W.J.; ARCULUS, R.J.; ELLIS, D.J.; FROST, C.D. A geochemical classification for granitic rocks. J. Petrology, 42, p. 2033-2048. 2001.

15. JORDT-EVANGELISTA, H. Igneous chamockites in the southeastem transition zone between the São Francisco Craton and the costeiro mobile belt, Brazil. Rev. Brasileira de Geociências. 26(2). p. 93-102. 1997.

16. LASA ENGENHARIA E PROSPECÇÕES S.A. Final report on the survey and processing of magnetometric and gamma-spectrometric data. Aerogeophysical Survey of Minas Gerais, Area 2. Pitangui - São João Del Rey - Ipatinga Project. 2001.

17. LAURENT, O.; MARTIN, H.; MOYEN, J.F.; DOUCELANCE, R. The diversity and evolution of late-Archean granitoids: Evidence for the onset of "modem-style" plate tectonics between 3.0 and 2.5 Ga. Lithos. 205. p. 208-235. 2014.

18. MACHADO FILHO, L.; RIBEIRO, M.W.; GONZALES, S.R.; SCHENINI, C.A.; SANTOS NETO, A.; PALMEIRA, R.C.B.; PIRES, J.L.; TEIXEIRA, W.; CASTRO, H.E.F. Geology of Sheets SF-23/24, Rio de Janeiro/Vitória. RADAMBRASIL Project, p. 27-304. 1983.

19. MACHADO, N.; NOCE, C.M.; LADEIRA, E.A.; BELO DE OLIVEIRA, O.A. U-Pb geochronology of Archean magmatism and Proterozoic metamorphism in the Quadrilátero Ferrifero, Southern Sao Francisco craton, Brazil. Geological Society of America Bulletin. 104. p.1221-1227. 1992.

20. MACHADO, N.; SCHRANK, A.; NOCE, C.M.; GAUTHIER, G. Ages of detrital zircon from Archean-Palaeoproterozoic sequences: implications for greenstone belt setting and evolution of a Transamazonian foreland basin in Quadrilátero Ferrifero, southeast Brazil. Earth and Planetary Science Letters. 141. p. 259-276. 1996.

21. MACKENZIE, W.S.; DONALDSON, C.H.; GUILDFORD, C. Atlas of igneous rocks and their textures.

London: Longford. 148p. 1982.

22. MANIAR, P.D. & PICCOLI, P.M. Tectonic discriminations of granitoids. Geological Society of America Bulletin, 101: p. 635-643. 1989.

23. MARSHALL, B. & SCHONWANDT, H.K. An Archaean sill complex and associated supracrustal rocks, Arveprinsen Ejland, north-east Disko Bugt, West Greenland. Geology of Greenland Survey Bulletin. 181. p. 87-102. 1999.

24. MACDONOUGH, W.F. & SUN, S.S. The composition of the Earth. Chemical Geology 120 (34). p. 223-253. 1995.

25. MIRANDA, D.A. & CHAVES, A.O. U-Th-Pb Age of Monazite and Phosphate-Silicate Reaction Cores from a Neoarchean Metachamoenderite in the southern São Francisco Craton (MG). (Article submitted for publication). 25 p. 2015.

26. NOCE, C.M. Geochronology of magmatic, sedimentary and metamorphic events in the Iron Quadrangle region, Minas Gerais. PhD Thesis. Institute of Geosciences, University of São Paulo, São Paulo. 128 p. 1995.

27. NOCE, C.M.; MACHADO, N.; TEIXEIRA, W. U-Pb Geochronology of gneisses and granitoids in the Quadrilatero Femfero (Southern São Francisco Craton): age constraints for Archean and Paleoproterozoic magmatism and metamorphism. Brazilian Journal of Geosciences. 28. p. 95-102. 1998.

28. NOCE, C.M.; TEIXEIRA, W.; QUÉMÉNEUR, J.J.G.; MARTINS, V.T.S.; BOLZACHINI, E. Isotope signatures of Palaeoproterozoic granitoids from the Southern São Francisco Craton and implications for the evolution of the Transamazonian Orogeny. Journal of South American Earth Sciences. 13. p. 225-239. 2000.

29. PACIULLO, F.V.P.; RIBEIRO, A.; CHERMAN, A.; CHRISPIM, S.; MACIEL, R.R. Folha Nepomuceno - SF.23-V-D-III, scale 1:100.000: explanatory note. Rio de Janeiro. 92p. (Geology of Brazil Programme: Geological Maps). 2007.

30. PERCIVAL, J.A.; SANBORN-BARRIE, M.; SKULSKI, T.; STOTT, G.M.; HELMSTAEDT, H.; WEUTE, D.J. Tectonic evolution of the western Superior Province from NATMAP and lithoprobe studies. Canadian Journal of Earth Sciences. 43. p. 1085-1117. 2006.

31. PINESE, J.P.P. Geochemistry, Isotopic Geology and Petrological Aspects of the Precambrian Mafic Dikes of the Lavras Region (MG), Southern Portion of the São Francisco Craton. PhD Thesis, Inst. of Geosciences, University of São Paulo, São Paulo, 178 p., 1997.

32. QUÉMÉNEUR J.J.G.; RIBEIRO, A.; TROUW, R.A.J.; PACIULLO, F.V.P.; HEILBRON, M. Geology of Folha Lavras. In: Companhia Mineradora de Minas Gerais. Geology and Mineral Resources of Southeastern Minas Gerais. South of Minas Project - Stage I. Final report. Belo Horizonte, COMIG/UFMG/UFRJ/UERJ, v. 1, p. 259-319. CD-ROM. 2003.

33. RICKWOOD, P.C. Boundary lines within petrologic diagrams which use oxides of major and minor elements. Lithos 22, p. 247-263. 1989.

34. ROMANO, R.; LANA, C.; ALKMIM, F.F.; STEVENS, G.; ARMSTRONG, R. Stabilisation of the Southern portion of the São Francisco Craton, SE Brazil, through a long-lived period of potassic magmatism. Precambrian Research 224, p. 143-159. 2013.

35. SMITHIES, R.H.; van KRANENDONK, M.J.; CHAMPION, D.C. The Mesoarchean emergence of modem-style subduction. Gondwana Research 11(1-2). p. 50-68. 2007.

36. STRECKEISEN A.L. Classification and Nomenclature of Plutonic Rocks. Recommendations of the IUGS Subcommission on the Systematics of Igneous Rocks. Geologische Rundschau. Internationale Zeitschrift fur Geologie. Stuttgart. Vol.63. p.773-85. 1974 A.

37. STRECKEISEN, A. How should chamockitic rocks be named? In: BELLIÈRE, J., DUCHESNE, J.C. (Eds.), Geologie des Domaines Cristallins. Vol. Centenary Sociéte Gèologique de Belgique, Liège. p. 349-360. 1974 B.

38. SUN, S.S. & MACDONOUGH, W.F. Chemical and isotopic systematics of oceanic basalts: implications for mantle composition and processes. In: SAUNDERS, A.D. & NORRY, M.J. (editors), Magmatism in the Ocean Basins. London. Geological Society Special Publication. p. 313-345. 1989.

39. TEIXEIRA, W. & FIGUEIREDO, M.C.H. An outline of Early Proterozoic crustal evolution in the São Francisco region, Brazil: a review. Precambrian Research. 53. p. 1-22. 1991.

40. TEIXEIRA, W.; CARNEIRO, MA; NOCE, C.M.; MACHADO, N.; SATO, K.; TAYLOR, P.N. Pb, Sr and Nd Interpretation of the Itabuna Alkaline Suite, São Francisco Craton, Brazil. Journal of South America Earth Sciences, 10 (1). p. 91-98. 1996.

41. TEIXEIRA, W.; SABATÉ, P.; BARBOSA, J.; NOCE, C.M.; CARNEIRO, M.A. Archean and Palaeoproterozoic tectonic evolution of the São Francisco Craton, Brazil. In: CORDANI, U.G.; MILANI, E.J.; THOMAZ-FILHO, A.; CAMPOS, D.A. (eds.) Tectonic evolution of South America. 31st International Geological Congress. Rio de Janeiro, p. 101-137. 2000.

42. TEIXEIRA, W.; ÁVILA, C.A.; CORDANI, U.G.; MARTINS, V.T.S.; VALENÇA, J.G. Isotopic data (U/Pb, Pb/Pb, Sm/Nd, Rb/Sr) from the Palaeoproterozoic plutonism of the Mineiro Belt, southern portion of the São Francisco Craton: tectonic implications. In: Symposium on the São Francisco Craton, 3, Salvador. SBG Annals. 1. p. 174-177. 2005.

43. VAN KRANENDONK, M.J.; SMITHIES, H.R.; HICKMAN, A.H.; CHAMPION, D.C. Review: secular tectonic evolution of Archean continental crust: interplay between horizontal and vertical processes in the formation of the Pilbara Craton, Australia. Terra Nova. 19(1). p. 1- 38. 2007.

44. WINDLEY, B.F. The Evolving Continents, 3rd ed. Chester, John Wiley, 526 p. 1995.

45. YARDLEY, B.W.D. Introduction to metamorphic petrology (Translated by R.A. Fuck). Ed. UNB. 340p. 1994.

46. ZHOU, J. & LI, X. GeoPlot: an excel VBA programme for geochemical data plotting. Computers & Geosciences 32. p. 554-560. 2006.

CHAPTER 2

U-Th-Pb AGE OF MONAZITE AND PHOSPHATE-SILICATE REACTION CROPS FROM A NEOARCHEAN METACHARNOENDERBITE IN THE SOUTH OF THE SÃO FRANCISCO CRATON (MG)

Daniel Andrade MIRANDA[1], Alexandre de Oliveira CHAVES[1]
Accepted for publication in Geosciences = Geosciences, Rio Claro, SP, Brazil - eISSN: 1980-900X

INTRODUCTION

Monazite crystals are phosphates of light rare earth elements (LREEs) and are found as an accessory mineral in various lithotypes with a meta- to peraluminous composition, including granitoids and granitic pegmatites, as well as metamorphic rocks with paraderivatives ranging from the green schist facies to the granulite facies (Overstreet, 1967). They usually contain U and Th, as well as essentially radiogenic Pb, derived from radioactive decay of the other two elements. There is negligible incorporation of common Pb in the natural growth of their crystals (Parrish, 1990), eliminating the need for robust isotopic correction for it in geochronological studies.

Since the 1990s, this mineral has become recognised as important in defining not only the age of magmatic crystallisation, but also metamorphism and deformation events, ages obtained through non-destructive, precise and high spatial resolution analyses of U, Th and Pb by electron microprobe (Suzuki & Adachi, 1991; Montei *et al,* 1996; Williams et al., 1999; Foster et al., 2004; Pyle et al., 2005; Vlach, 2010), which are equivalent to the isotopic ages obtained by the U-Pb method. This equivalence is confirmed by several papers in the literature. For example, in their article, Dahl et *al.* (2005) validate the accuracy and precision of the U-Th-Pb ages obtained by electron microprobe in monazite from granite-gneiss and metapelitic rocks from Precambrian polymetamorphic terrains, by comparing them with U-Pb isotopic ages obtained by SHRIMP (Sensztzve *High Resolution lon MicroProbe)* in the edge and core portions of the crystals. Chaves et *al.* (2013) also demonstrate that the U-Th-Pb ages between 505-580 Ma obtained by electron microprobe in detrital monazite crystals from marine *placers*, which are useful for defining the age of source areas that provide material to feed sedimentary basins, are perfectly comparable to the U-Pb isotopic ages between 530-580 Ma obtained by LA-ICP- MS (Laser *Ablation Inductively Coupled Plasma Mass Spectrometry).*

In the absence of metamorphic fluids, monazite is believed to remain a closed system under conditions of medium to high metamorphic grade. However, in the presence of silicic fluids, monazite often reacts to form allanite and/or apatite under variable metamorphic facies temperature and pressure conditions (Smith & Barreiro, 1990; Ward *et al.,* 1991; Lanzirotti & Hanson, 1996). Finger *et al.* (1998) show that monazite crystals in granite-gneisses subjected to metamorphic fluids under amphibolite facies conditions become unstable and covered by apatite-allanite-epidote crowns.

The aim of this paper is to present the U-Th-Pbj chemical age (non-isotopic) of monazite crystals from an Archaean metachamoenderite (granodiorite with orthopyroxene, metamorphised) outcropping in the southern portion of the São Francisco Craton (MG) and to discuss the redistribution of elements during the partial replacement of magmatic monazite by crowns of phosphates and silicates formed during the metamorphic processes undergone by the rock. Geodynamic aspects related to the emplacement and metamorphism of the metachamoenderite are also addressed here.

REGIONAL GEOLOGICAL FRAMEWORK

The southern portion of the São Francisco Craton (CSFM) is made up of granite-gneiss-migmatite terrains and Archaean *greenstone belts* (Rio das Velhas Supergroup), as well as Palaeoproterozoic clastic-chemical metasedimentary rocks (including itabirites in the Iron Quadrangle) of the Minas Supergroup and Neoproterozoic pelite-carbonate sedimentary covers of the Bambuí Group (Figure 1). In these Archaean terrains there are three main metamorphic complexes, called Belo Horizonte, Bonfim and Campo Belo. Most of them consist of gneisses and migmatites of tonalite-trondhjemite-granodiorite (TTG) composition, whose protoliths were formed between 3200 and 2900 Ma, as well as calcium-alkaline syntectonic granitoids formed in the Rio das Velhas Orogeny, with ages around 2780 Ma, post-tectonic granitoids formed between 2700 and 2600 Ma, as well as mafic and mafic-ultramafic intrusions, are found in the middle of the TTG terrains (Noce et *al.,* 1998; Teixeira et *al.,* 2000).

In regional terms, the evolution of the South Minas Gerais Palaeoproterozoic Province (Teixeira et *al.,* 2005), a domain of the action of metamorphism and deformation on the archaean rocks of the CSFM, should be highlighted. It includes the Mineiro Belt at the extreme SE edge of the CSFM, a marker of Palaeoproterozoic events of an accretionary and ensialic nature. This belt, with a general SW-NE direction and a SE to NW tectonic gradient, holds the initial record of an intra-oceanic island arc from around 2300 Ma, added to an Andean-type continental margin. A magmatic arc would have been built on this margin, with plutons from 2200 Ma, until the moment of the continental collision at 2100 Ma, when the sediments of the Minas Supergroup were deformed in the Quadrilátero Ferrífero region (Teixeira & Figueiredo, 1991; Machado et *al.,* 1992; Ávila et *al.,* 2010; Noce et *al.*, 2000; Alkmim, 2004).

According to Carneiro & Barbosa (2008), the sialic crustal segment of the CSFM, in which the study area is located, is made up predominantly of gneisses, granitoids, amphibolites, mafic and ultramafic rocks and chamockites, which were initially grouped by Machado Filho *et al.* (1983) into the Divinópolis and Barbacena metamorphic complexes. However, Teixeira et *al.* (1996) united the Divinópolis and Barbacena metamorphic complexes into a single complex, which they called the Campo Belo Metamorphic Complex (CMCB), whose geological evolution began in the Mesoarchean. Supracrustal relicts occur locally, which may correspond to Archaean sequences and/or Palaeoproterozoic sequences correlating to the Iron Quadrangle (Teixeira et *al.,* 2000). With regard to metamorphism, most of the rocks in the CMCB were generated or subjected to metamorphic conditions that reached the amphibolite and granulite facies (Femandes & Carneiro, 2000).

This whole complex is cut by mafic dykes (diabase and amphibolite) of various generations, as described by Chaves (2013).

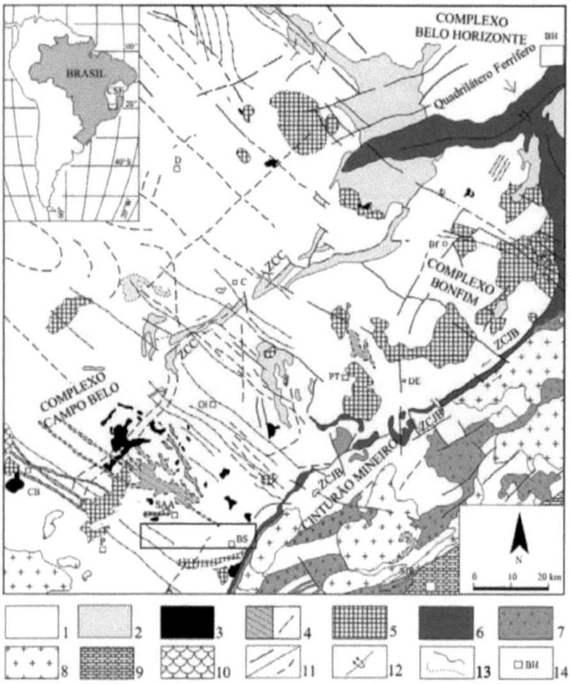

Figure 1 - Schematic geological map of the southern portion of the São Francisco Craton showing the main Archaean and Palaeoproterozoic plutonic bodies (Extracted from Campos *et al.*, 2003). 1 - Archaean basement partially reworked in the Palaeoproterozoic. - Rio das Velhas *Greenstone*. 3 - Neoarchean plutonic ultramafic suites. - Neoarchean to Mesoproterozoic gabbros and diorites. 5 - Neoarchean granitoids. 6 - Minas Supergroup. 7 - Indiscriminate *greenstone* sequences. 8 - Palaeoproterozoic granitoids of the Minas Belt. 9 - São João Del Rey/Andrelândia Groups. 10 - Undivided Neoproterozoic cratonic cover - Bambuí Group. 11 - Faults and fractures (ZCC = Cláudio Shear Zone; ZCJB = Jeceaba-Bom Sucesso Shear Zone). 12 - Axial planes. 13 - Lithological contacts. 14 - Towns: BH = Belo Horizonte, Bf = Bonfim, BS = Bom Sucesso, CB = Campo Belo, C = Cláudio, DE = Desterro de Entre Rios, D = Divinópolis, O1 = Oliveira, P = Perdões, PT = Passa Tempo, SAA = Santo Antônio do Amparo, SJR = São João Del Rei. The rectangle next to the town of Bom Sucesso represents the image in Figure 2.

The metachamoenderite studied, belonging to the Chamockitic Complex of Babylon, described by Quéméneur *et al.* (2003), outcrops in the domain to the west of the Serra de Bom Sucesso (Figure 2). This domain contains mafic and ultramafic neoarchean rocks, as well as granites and bodies that vary from chamockite to enderite. According to Pinese (1997), local gabbro-norite bodies are dated at 2658 ± 44 Ma (Sm-Nd isochron of total rock and mineral concentrates), which he interprets as the age of intrusion.

Paciullo et *al.* (2007) use the name Rio do Amparo for the metalluminous chamockite rocks. They found an age close to 2718 ± 13 Ma (U-Pb zircon and contact relations with adjacent rocks) and a Sm-NdiDM model age of 2768 Ma (total rock) for these chamockites, which are gneissified. Quéméneur (1995) published an age of 2660 ± 30 Ma (Rb-Sr isochron, whole rock) for these same rocks. Paciullo et *al.* (2007) interpret the model age to represent the separation from the protolith, the U-Pb age to represent the crystallisation of the chamockites and the Rb-Sr age to represent the opening in the Rb-Sr system at the end of metamorphism.

The Bom Sucesso granitoid not only has petrographic and geochemical similarities with the chamockitic Rio do Amparo rocks, but also geochronological similarities, with a magmatic crystallisation age of 2708 Ma (Rb-Sr isochrone, Quéméneur et *al.,* 2003). Like the chamockites, it is gneissified.

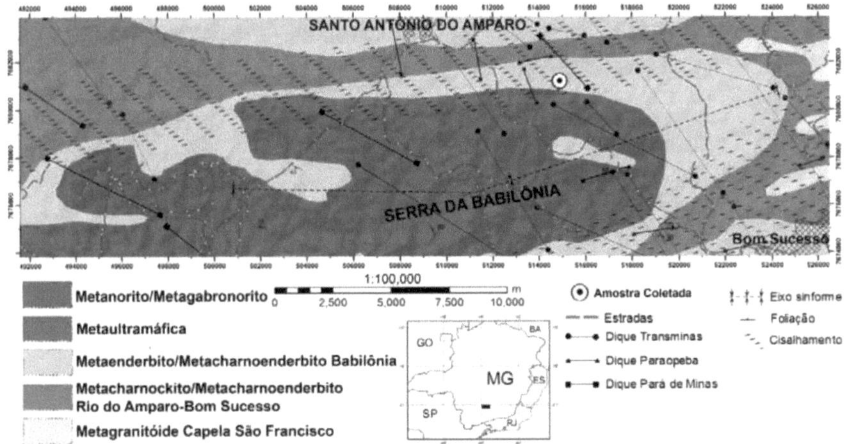

Figure 2. Detail of the study area, showing where the analysed rock was collected.

MATERIALS AND METHODS

The sample of metachamoenderite studied was collected and described during the field stage and taken to the laboratories of the Geosciences Institute of the Federal University of Minas Gerais (IGC/UFMG), where the rock was sliced for petrographic studies under a microscope. Using the polished thin sheet investigated in the petrographic studies, the chemical compositions of the edge and centre portions of six 50 to 250 pm monazite crystals were obtained, with a total of 18 analyses obtained at the UFMG Microscopy Centre (CM-UFMG) on a JEOL microprobe, model JXA-8900, using wavelength dispersive spectrometry (WDS). The monazites were imaged using the backscattered electron technique on this microprobe.

To obtain geochronological data, the U-Th-Pbf chemical dating method for monazite developed by Chaves *et al.* (2013) was used. The measurement conditions and analytical standards used are shown in Table 1. X-ray peak overlaps between Y and Pb did not need to be corrected because there was no measurement in PbMa (Lead M alpha), but only in PbMb (Lead M beta). However, in order to avoid errors in the ages obtained, the interference of ThMz (Thorium M gamma) on the UMb (Uranium M beta) measured had to be corrected, following Scherrer et *al.* (2000) in adaptation to the conditions at CM-UFMG, as follows:

$$U\ corrected = Moist - (0.006365\ x\ Th\ measured).$$

The U-Th-Pbr point chemical ages were calculated using the *EPMA Dating software* (Pommier *et al.,* 2004), using the relationship:

$$Pb = \left\{Th \times [\exp(\lambda_{232}t) - 1] \times \frac{M_{208}}{M_{232}}\right\} + \left\{U \times [\exp(\lambda_{238}t) - 1] \times \frac{M_{206}}{M_{238}} \times 0.9928\right\}$$
$$+ \left\{U \times [\exp(\lambda_{235}t) - 1] \times \frac{M_{207}}{M_{235}} \times 0.0072\right\}$$

Onde:

Pb, Th e U(corrigido) são as concentrações em ppm obtidas na microssonda. *t* é a idade em milhões de anos (Ma)

M_{206}, M_{207}, M_{208}, M_{232}, M_{235}, M_{238} são as massas atômicas de ^{206}Pb, ^{207}Pb, ^{208}Pb, ^{232}Th, ^{235}U, ^{238}U.

$$\lambda_{232} = 0.49475 \times 10^{-4}\ Ma^{-1}$$
$$\lambda_{235} = 9.8485 \times 10^{-4}\ Ma^{-1}$$
$$\lambda_{238} = 1.55125 \times 10^{-4}\ Ma^{-1}$$

$$\frac{^{238}U}{^{235}U + ^{238}U} = 0.9928 \qquad \frac{^{235}U}{^{235}U + ^{238}U} = 0.0072$$

The average ages and their statistical parameters were calculated using *Isoplot software* (Ludwig, 2003).

Table 1. Measurement conditions and analytical standards used in the quantitative investigation of the elements present monazites using the CM-UFMG electron microprobe (Chaves *et al.*, 2013).

Method = quantitative WDS (wavelength dispersive spectroscopy)					
Acceleration voltage = 25kv					
Current strength = 50nA					
Electron beam diameter = 2 micrometres					
			Counting time (sec.)		
Element	**X-ray**	**Crystal**	**Pico**	**Background**	**Analytical Standard**
Y	La	TAP	20,0	10,0	YAG (Ytrium Aluminium Gamet)
Dy	Lb	LIF	20,0	10,0	$DyPO_4$
P	Ka	PETJ	10,0	5,0	monazite
Yes	Ka	TAP	10,0	5,0	$ThSiO_4$ (synthetic thorite)
Gd	Lb	LIF	20,0	10,0	$GdPO_4$
Pb	Mb	PETJ	200,0	100,0	crocoite
Th	Ma	PETJ	20,0	10,0	$ThSiO_4$
Sm	Lb	LIF	20,0	10,0	REE2
U	Mb	PETJ	150,0	75,0	OU_2 (summarised)
Ca	Ka	PETJ	10,0	5,0	$Ca_2P_2O_7$ (synthetic apatite)
Nd	Lb	LIF	20,0	10,0	$NdPO_4$
La	La	PETJ	10,0	5,0	monazite
Pr	Lb	LIF	20,0	10,0	$PrPO_4$
Ce	La	LIF	10,0	5,0	monazite

RESULTS

PETROGRAPHIC DESCRIPTION

The rock collected is mesocratic, equigranular, medium-grained and foliated. Microscopically, the rock described has a granolepidoblastic texture with orientated biotite crystals marking the foliation (Figure 3A and B). Its average composition is 35 to 40% quartz, very recrystallised, with undulating extinction and contacts varying from interlobed to polygonal, 35% plagioclase, 15% biotite, 10% potassium feldspar and 5%

orthopyroxene (orthopyroxene is replaced by a mixture of amphibole, biotite and quartz - Figure 3C). Accessory minerals include ilmenite, titanite, apatite and monazite. Monazite has alteration crowns for secondary apatite, allanite and epidote (Figure 4, A and B).

Like many regional lithotypes, the rock investigated crystallised/recrystallised under a high metamorphic degree in granulite facies conditions, and was probably retrometamorphised while still in the amphibolite facies, due to the verification of the amphibole generation process replacing orthopyroxene.

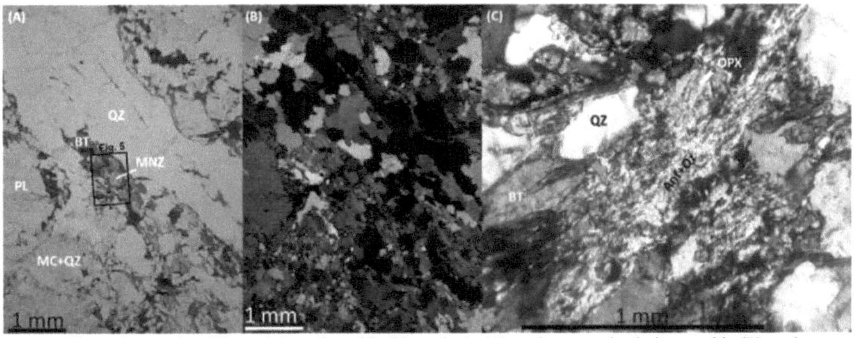

Figure 3 - Photomicrographs of the rock investigated in this study. In (A), under crossed polarisers and in (B), under crossed polarisers, it is possible to see the granolepidoblastic texture of the slide and the foliation marked by biotite. (C) Detail of orthopyroxene in the slide, replaced by a mixture of amphibole, biotite and quartz (crossed polarisers). QZ = Quartz, PL = Plagioclase, BT = Biotite, MC = Microcline, MNZ = Monazite, OPX = Orthopyroxene.

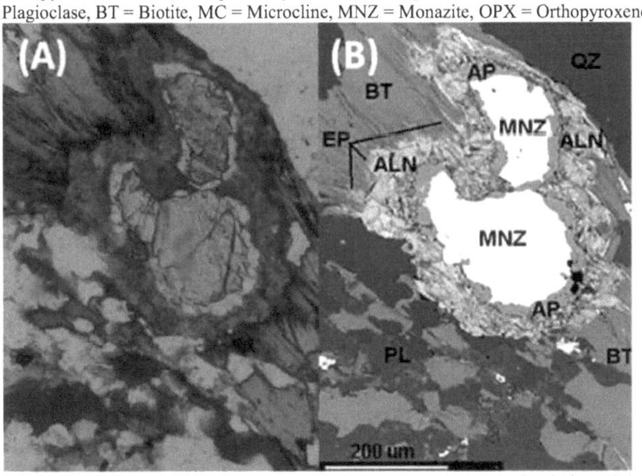

Figure 4 - Detail photomicrograph of a monazite crystal under crossed polarisers (A) and its respective backscattered electron image (B). MNZ = Monazite, AP = Apatite, ALN = Allanite, EP = Epidote, BT = Biotite, QZ = Quartz, PL = Plagioclase.

MINERAL CHEMISTRY

The monazite crystals investigated, Cia C6, are anhedral to subhedral and sometimes intemely homogeneous, without zoning, sometimes heterogeneous, with zoning features, with lighter portions being slightly richer in Th (Figure 5). Three analyses were carried out per crystal (A, B and C) and the chemical composition of the points analysed (Table 2), plotted on Linthout's (2007) classification diagram, indicates that all the crystals plot in the monazite field (Figure 6), all being chemically very similar. The higher Ce content compared to the other ETRs shows that the crystals are monazite-(Ce) with ThCh contents between 3.10 and 5.63 %, UO2 between 0.08 and 0.13 % and PbO between 0.40 and 0.73 %. The ETR graphs normalised to the chondrite (Sun & MacDonough, 1989) show that all the crystals are extremely similar in terms of the pattern of behaviour of these elements (Figure 5), which suggests that these crystals are co-genetic.

The crowns found around the monazite crystals in figure 4 are confirmed to be apatite, allanite and epidote by means of the energy dispersive spectra (EDS) shown in figure 7.

GEOCRONOLOGY

Table 3 shows the point ages and contents of U, Th and Pb, and their respective errors, obtained from the six monazite crystals investigated. The average of all the ages shown in Table 3 produces a value of 2657 ± 36 Ma (Figure 8).

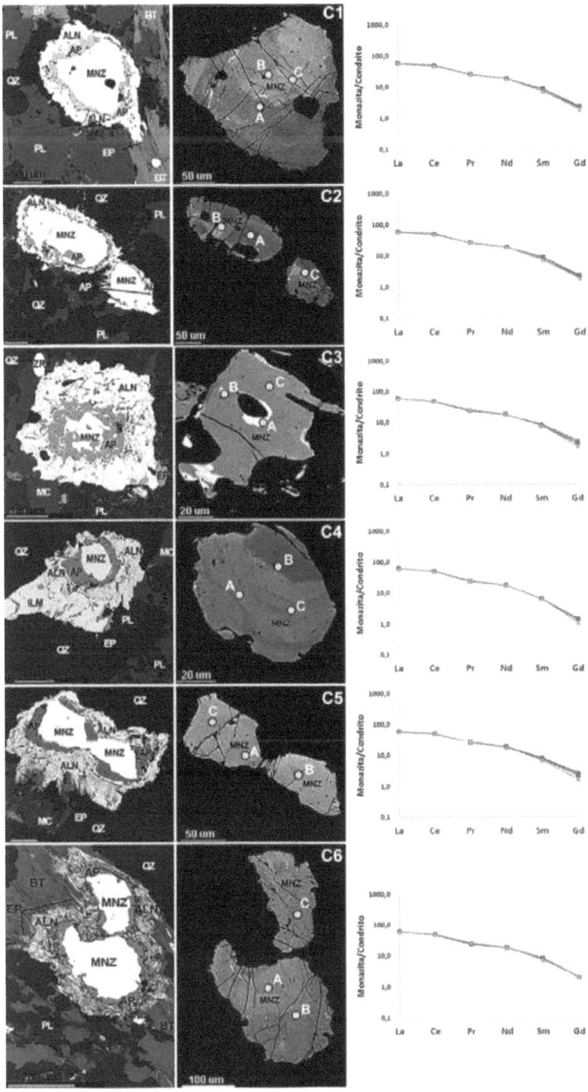

Figura 5. Backscattered electron images of the monazite crystals analysed and ETR graphs normalised to the chondrite (Sun & MacDonough, 1989). The dots represent the electron microprobe analyses listed in Table . MNZ = Monazite, AP = Apatite, ALN = Allanite, EP = Epidote, ILM = Ilmenite, BT = Biotite, QZ = Quartz, PL = Plagioclase, MC = Microcline.

Table 2. Chemical composition of the monazite crystals from the investigated metachamoenderite (oxides in weight %).

	Y_2O_3	Dy_2O_3	P2O5	SiO_2	Gd_2O_3	PbO	ThO_2	Sm_2O_3	UO_2	CaO	Nd_2O_3	La_2O_3	Pr2O3	Ce_2O_3	Total
Cl-A	0,23	0,16	27,39	0,99	0,50	0,46	3,73	1,26	0,08	0,32	10,21	16,73	2,88	36,26	101,20
Cl-B	0,33	0,21	27,28	1,40	0,54	0,67	5,43	1,54	0,12	0,46	10,05	16,08	2,86	34,73	101,69
Cl-C	0,25	0,16	27,31	1,22	0,44	0,70	5,54	1,30	0,13	0,47	9,93	16,61	2,89	34,89	101,84
C2-A	0,26	0,16	27,73	1,03	0,54	0,48	3,67	1,31	0,11	0,32	10,26	16,89	2,90	35,75	101,39
C2-B	0,27	0,14	27,01	1,41	0,49	0,73	5,55	1,42	0,12	0,40	10,43	16,44	2,88	34,32	101,62
C2-C	0,28	0,09	27,52	1,15	0,54	0,52	4,06	1,44	0,09	0,34	10,26	16,52	2,96	35,66	101,44

C3-A	0,32	0,19	27,18	1,30	0,61	0,69	5,47	1,53	0,10	0,48	9,42	16,49	2,80	34,73	101,30
C3-B	0,32	0,08	27,30	1,29	0,50	0,67	5,25	1,38	0,12	0,45	9,81	16,50	2,62	34,91	101,21
C3-C	0.27	0,11	27,54	1,19	0,41	0,62	4,90	1,40	0,11	0,45	9,88	16,76	2,92	35,32	101,88
C4-A	0,19	0,07	27,13	1,18	0,33	0,55	4,23	1,12	0,09	0,29	9,66	17,17	2,78	36,51	101,31
C4-B	0,20	0,15	27,18	1,08	0,33	0,48	3,73	1,15	0,09	0,26	9,65	17,43	2,68	36,09	100,48
C4-C	0,14	0,10	27,16	1,17	0,26	0,53	4,07	1,16	0,08	0,31	9,49	17,29	2,83	35,67	100,25
C5-A	0,31	0,00	27,38	1,20	0,65	0,58	4,53	1,41	0,13	0,44	10,45	16,32	2,97	35,17	101,53
C5-B	0,31	015	26,85	1,20	0,55	0,60	4,72	1,40	0,12	0,43	10,23	16,30	2,90	35,32	101,06
C5-C	0,30	0,12	27,26	0,91	0,41	0,41	3,14	1,25	0,08	0,35	9,80	16,84	2,83	36,80	100,48
C6-A	0,31	0,09	27,00	1,39	0,49	0,70	5,63	1,46	0,12	0,47	9,96	16,53	2,89	34,65	101,69
C6-B	0,31	0,03	27,25	1,32	0,47	0,69	5,49	1,41	0,11	0,46	10,13	16,36	2,67	34,24	100,91
C6-C	0,28	0,11	27,76	0,87	0,49	0,40	3,10	1,26	0,09	0,52	10,01	16,70	2,89	35,35	99,81

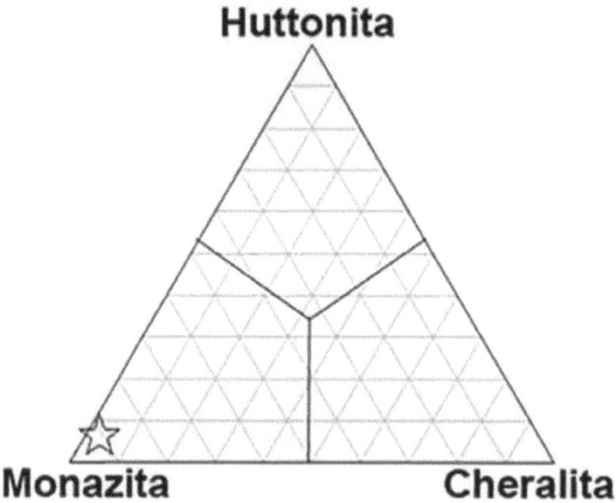

Figura 6. Ternary diagram $2REEPO_4$ - $CaTh(PO_{(4)})_{(2)}$ - $2ThSiO_{(4)}$ (Linthout, 2007). The star represents the composition of the 18 points analysed.

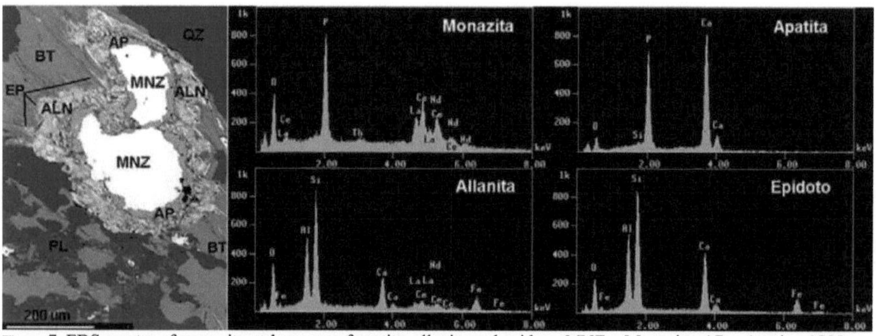

Figure 7. EDS spectra of monazite and crowns of apatite, allanite and epidote. MNZ = Monazite, AP = Apatite, ALN = Allanite, EP = Epidote, BT = Biotite, QZ = Quartz, PL = Plagioclase.

Table 3. Point ages and U, Th and Pb contents (and their respective errors) obtained from monazite crystals. M Pb is the average atomic mass of lead, calculated according to Pommier *et al.* (2004).

	Age Ma	Error Age Ma	U ppm	Error U ppm	Th PPm	Error Th ppm	Pb PPm	Error Pb ppm	M Pb
Cl-A	2619	178	452	150	32806	656	4298	150	207,9
Cl-B	2605	137	728	150	47701	954	6247	150	207,9

C1-C	2630	135	792	150	48703	974	6470	150	207,9
C2-A	2660	176	738	150	32234	645	4447	150	207,8
C2-B	2736	138	774	150	48782	976	6758	150	207,9
C2-C	2689	168	593	150	35688	714	4864	150	207,9
C3-A	2684	140	584	150	48053	961	6424	150	207,9
C3-B	2674	141	738	150	46155	923	6238	150	207,9
C3-C	2634	147	678	150	43061	861	5718	150	207,9
C4-A	2704	164	574	150	37191	744	5078	150	207,9
C4-B	2668	178	550	150	32744	655	4428	150	207,9
C4-C	2724	170	487	150	35741	715	4883	150	207,9
C5-A	2621	151	848	150	39836	797	5375	150	207,8
C5-B	2607	148	812	150	41453	829	5523	150	207,8
C5-C	2679	201	494	150	27629	553	3769	150	207,8
C6-A	2626	134	761	150	49433	989	6535	150	207,9
C6-B	2664	138	645	150	48220	964	6424	150	207,9
C6-C	2614	198	594	150	27208	544	3667	150	207,8

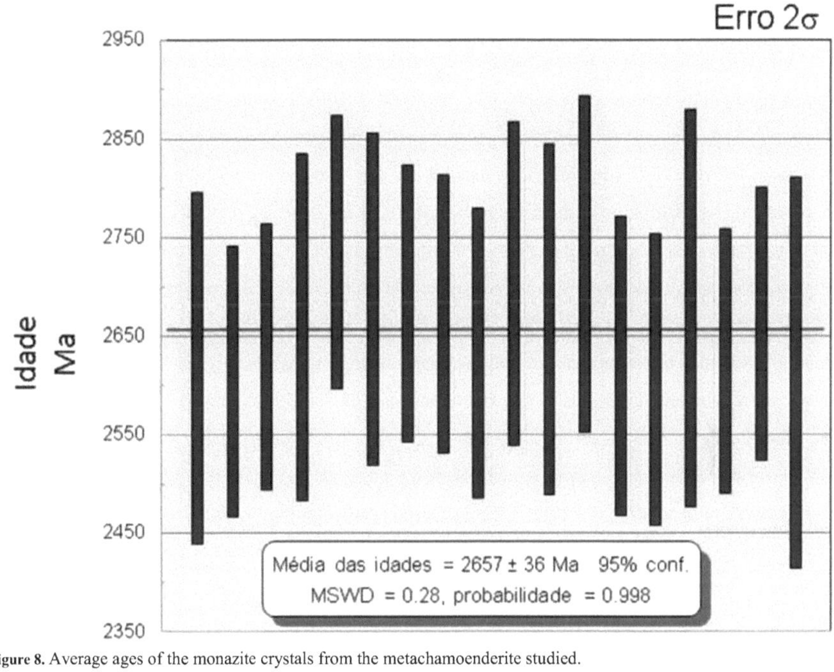

Figure 8. Average ages of the monazite crystals from the metachamoenderite studied.

DISCUSSION AND CONCLUSIONS

Monazite is very useful for dating high temperature/pressure magmatic and metamorphic events (Parrish, 1990). The typical high concentrations of U and Th in monazite also contribute to the reliability of the ages found. As a result, monazite often produces robust U-Th-Pb ages. In geochronological terms, as mentioned in the Regional Geology section of this article, Paciullo *et al.* (2007) interpret the Sm-NdiDM model age of 2768 Ma (total rock) of the local metalluminous metachamockite rocks as representing the age of separation from the protolith, the U-Pb age (zircon) of 2718 ± 13 Ma as being close to the magmatic crystallisation of the chamockites and the age of 2660 ± 30 Ma (Rb-Sr isochrone, total rock, Quéméneur, 1995) as the moment of opening of the Rb-Sr system at the end of metamorphism. The Bom Sucesso granitoid, with a magmatic crystallisation age of 2708 ± 5 Ma (Quéméneur et *al.,* 2003), would also be part of the same geodynamic context.

Specifically with regard to monazite, since its crystals can be extremely homogeneous or heterogeneous depending on their geological history, the backscattered electron images generated by the microprobe can provide thermotectonic information about the

crystal growth and recrystallisation patterns related to metamorphic reactions (Williams *et al.,* 2007). In the backscattered electron images in figure 5, crystals such as C4 apparently have a typical magmatic zoning pattern, with two ages corresponding to the oldest in table3. Cl, for example, shows an irregular zoning pattern, apparently related to metamorphic recrystallisation processes, with redistribution of Th and slight loss of radiogenic Pb by the crystal during recrystallisation, as observed in polymetamorphic rocks from Antarctica (Black et al., 1984), which would lead to younger ages, similar to those of the Cl crystal in table 3.

The average age of 2657 ± 36 Ma (Figure 8) found for the monazite crystals of the metachamoenderite investigated here points to contemporaneity with the 2660 ± 30 Ma metamorphism event suggested by Quéméneur (1995) for this context. However, it is suggested that the ages corresponding to the oldest in table 3 (such as those of crystal C4 in figure 5) are related to the age of magmatic crystallisation of the monazites studied, placing the crystallisation of the metachamoenderite in the same age interval (2718-2708 Ma) as the adjacent metachamockites. The temporal and spatial similarity between all these rocks refers to a mantle-crust interaction environment related to the late-collisional period of the Rio das Velhas event, with magmatism and metamorphism developing under pressure and temperature conditions typical of granulite facies. These conditions are attested to by the presence of orthopyroxene in the metachamoenderite investigated, which was also subjected to amphibolite facies retrometamorphism, given the replacement of orthopyroxene by amphibole and the appearance of biotite.

With regard to the crown-shaped textures around the monazite crystals, they reveal that this mineral present in the rock studied did not remain a totally closed system during the regional metamorphism from granulite to amphibolite facies. The consumption of monazite suggests the presence of metamorphic fluids that would have carried Ca, Fe, Si and Al to the limits of its crystals. However, as the alteration textures are concentric, the kinematics of the reactions were apparently controlled by the diffusion of elements from the monazite to the crown minerals, and vice versa. The apatite formed was only able to grow along, or very close to, the structure of the original monazite. In this way, it seems that the PO4 tetrahedrons of the monazite served directly as the basis for the growth of the apatite. Because ETRs have a low diffusion coefficient in accessory minerals (Liang & Watson, 1995), it can be assumed that the ETRs released during the consumption of monazite diffused very slowly along the apatite zone and towards the allanite/epidote crowns (Figure 5 and Figure 7). In any case, the reaction rate decreased as the apatite zone grew. The growth of the apatite zone finally prevented the monazite from coming into contact with the eventual metamorphic fluid, which explains why the complete consumption of the monazite was never achieved.

Unlike the apatite that grew towards the centre of the monazite, the allanite and epidote crowns were formed as a "fringe" at the edges of the original monazite crystal. Allanite protrusions along the epidote zone suggest that the allanite was successively replacing the newly formed epidote crowns. The nucleation and growth of the ETR-rich allanite crown adjacent to the apatite crown was most likely driven by the particularly high concentration of ETR in this region. The drop in ETR availability away from the apatite zone resulted in the formation of a crown poorer in ETR, which is epidote. This observation supports the assertion that ETR are extremely immobile during metamorphism (Finger *et al.,* 1998).

It is likely that the Ca needed to form apatite and epidote-allanite came from the consumption of plagioclase. The Fe and Al needed to form epidote-allanite came from the consumption of biotite and ilmenite.

Considering an interval of around 40 Ma between the age of magmatic crystallisation and the final age of metamorphism at 2657 ±36 Ma for the rock investigated and the presence of phosphate-silicate reaction coronas in the monazites, it is suggested that the growth of the coronas would have started soon after the crystallisation of the metachamoenderite, around 2700 Ma, as there are reaction coronas around crystals that have an igneous growth pattern. The hydrated minerals in the reaction crowns continued to develop due to the action of hydrothermal fluids in the amphibolite facies retrometamorphism during the late-collisional extension of the Rio das Velhas orogeny until around 2660 Ma, given that they grew around crystals with an irregular pattern reminiscent of monazite recrystallisation features. Under these P-T conditions and the 40 Ma time interval considered, the diffusion of elements would not have been enough to completely consume the monazite.

ACKNOWLEDGEMENTS

To CNPq for the scholarship awarded to the authors. To the UFMG Geosciences Institute, which made the CPMTC laboratories available to support this research.

BIBLIOGRAPHICAL REFERENCES

1. ALKMIM, F.F. What makes a craton a craton? The São Francisco Craton and the Almeidian revelations in delimiting it. In: MANTESSO-NETO, V.; BARTORELLI, A.; DAL RÉ CARNEIRO C.; BRITO-NEVES, B.B. (Org.). Geology of the South American Continent. Evolution of the work of Fernando Flávio Marques de Almeida. Becca, São Paulo, p. 17-35, 2004.

2. ÁVILA, C.A.; TEIXEIRA, W.; CORDANI, U.G.; MOURA, C.A.V.; PEREIRA, R.M. Rhyacian (2.23-2.20 Ga) juvenile accretion in the southern São Francisco Craton, Brazil: Geochemical and isotopic evidence from the Serrinha magmatic suite, Mineiro belt. Journal of South American Earth Sciences, v. 29, p. 464-482, 2010.

3. BLACK, L.P.; FITZGERALD, J.D.; HARLEY, S.L. Pb isotopic composition, colour, and microstructure of monazites from a polymetamorphic rock in Antarctica. Contributions to Mineralogy and Petrology, v. 85, p. 141-148, 1984.

4. CAMPOS, J.C.S.; CARNEIRO, M.A.; BASEI, M.A.S. U-Pb evidence for Late Neoarchean crustal reworking in the southern São Francisco Craton (Minas Gerais, Brazil). Anais da Academia Brasileira de Ciências, v. 75, n. 4, p. 497-511, 2003.

5. CARNEIRO, M.A. & BARBOSA, M.S.C. Geological and tectonic implications of the magnetometric interpretation of the Oliveira region, Minas Gerais. Revista Brasileira de Geofísica, v. 26, n. 1, p. 87-98, 2008.

6. CHAVES, A.O. Mafic dyke swarms of Minas Gerais - the state of the art. Geonomos, v. 21, n. l,p. 29-33, 2013.

7. CHAVES, A.O.; OLIVEIRA, E.K.; GARCIA, L.R.A. Development of the U-Th-Pb chemical dating method of monazite by electron microprobe at UFMG. Geonomos, v. 21, n. 2, p. 13-18, 2013.

8. DAHL, P.S.; HAMILTON, M.A.; JERCINOVIC, M.J.; TERRY, M.P.; WILLIAMS, M.L.; FREI, R. Comparative isotopic and Chemical geochronometry of monazite, with implications for U-Th-Pb dating by electron microprobe: An example from metamorphic rocks of the eastem Wyoming Craton (U.S.A.). American Mineralogist, v. 90, p. 619-638, 2005.

9. FERNANDES, R.A. & CARNEIRO, M.A. The Campo Belo Metamorphic Complex (Southern São Francisco Craton): lithodemic units and tectonic evolution. Revista Brasileira de Geociências, v. 30, n. 4, p. 671-678, 2000.

10. FINGER, F.; BROSKA, L; ROBERTS, M.P.; SCHERMAIER, A. Replacement of primary monazite by apatite-allanite-epidote coronas in an amphibolites facies granite gneiss from the eastem Alps. American Mineralogist, v. 83, p. 248-258, 1998.

11. FOSTER, G.; PARRISH, R.R.; HORSTWOOD, M.S.; CHENERY, S.; PYLE, J.; GIBSON, H.D. The generation of prograde P-T-t points and paths; a textural, compositional, and chronological study of metamorphic monazite. Earth Planetary Science Letters, v. 228, p. 125142, 2004.

12. LANZIROTTI, A. & HANSON, G.N. Geochronology and geochemistry of multiple generations of monazite from the Wepawaug Schist, Connecticut, USA: Implications for monazite stability in metamorphic rocks. Contributions to Mineralogy and Petrology, v. 125, p. 332-340, 1996.

13. LAURENT, O.; MARTIN, H.; MOYEN, J.F.; DOUCELANCE, R. The diversity and evolution of late-Archean granitoids: Evidence for the onset of "modem-style" plate tectonics between 3.0 and 2.5 Ga.

Lithos, v. 205, p. 208-235, 2014.

14. LIANG, Y. & WATSON, E.B. A simple model for sector zoning in slowly grown crystals: Implications for growth rate and lattice diffusion, with emphasis on accessory minerals in crustal rocks. American Mineralogist, v. 80, p. 1179-1187, 1995.

15. LINTHOUT, K. Tripartite Division of THE System 2REEPO4 - CaTh(PO4)2 - 2ThSiO4, Discreditation of Brabantite, and Recognition of Cheralite as the name for members dominated by CaTh(PO4)2. Canadian Mineralogist, v. 45, p. 503-508, 2007.

16. LUDWIG, K.R. Isoplot/Ex 3.00: A geochronological toolkit for Microsoft Excel. Berkeley Geochronology Centre, Special Publication, n.4, 70 p. 2003.

17. MACHADO FILHO, L.; RIBEIRO, M.W.; GONZALES, S.R.; SCHENINI, C.A.; SANTOS NETO, A.; PALMEIRA, R.C.B.; PIRES, J.L.; TEIXEIRA, W.; CASTRO, H.E.F. Geology of Sheets SF-23/24, Rio de Janeiro/Vitória. Projeto RADAMBRASIL, p. 27-304, 1983.

18. MACHADO, N.; NOCE, C.M.; LADEIRA, E.A.; BELO DE OLIVEIRA, O.A. U-Pb geochronology of Archean magmatism and Proterozoic metamorphism in the Quadrilátero Ferrifero, Southern Sao Francisco craton, Brazil. Geological Society of America Bulletin, v. 104, p. 1221-1227, 1992.

19. MONTEL, J.; FORET, S.; VESCHAMBRE, M.; NICOLLET, C.; PROVOST, A. Electron microprobe dating of monazite. Chemical Geology, v. 131, p. 37-53, 1996.

20. NOCE, C.M.; MACHADO, N.; TEIXEIRA, W. U-Pb Geochronology of gneisses and granitoids in the Quadrilatero Femfero (Southern São Francisco Craton): age constraints for Archean and Paleoproterozoic magmatism and metamorphism. Revista Brasileira de Geociências, v. 28, p. 95-102, 1998.

21. NOCE, C.M.; TEIXEIRA, W.; QUÉMÉNEUR, J.J.G.; MARTINS, V.T.S.; BOLZACHINI, E. Isotope signatures of Palaeoproterozoic granitoids from the Southern São Francisco Craton and implications for the evolution of the Transamazonian Orogeny. Journal of South American Earth Sciences, v. 13, p. 225-239, 2000.

22. OVERSTREET, W.C. The geological occurrence of monazite. Geological Survey Professional Papers, v. 530, p. 327, 1967.

23. PACIULLO, F.V.P.; RIBEIRO, A.; CHERMAN, A.; CHRISPIM, S.; MACIEL, R.R. Geological map of

the Nepomuceno Sheet (SF.23-V-D-III) Scale 1:100.000. Explanatory note. Geology of Brazil Programme: Geological Maps, Rio de Janeiro, 92 p., 2007.

24. PARRISH, R.R. U-Pb dating of monazite and its applications to geological problems. Canadian Journal of Earth Science, v. 27, p. 1431-1450, 1990.

25. PINESE, J.P.P. Geochemistry, Isotopic Geology and Petrological Aspects of the Precambrian Mafic Dikes of the Lavras Region (MG), Southern Portion of the São Francisco Craton. PhD Thesis, Inst. of Geosciences, University of São Paulo, São Paulo, 178 p., 1997.

26. POMMIER, A.; COCHERIE, A.; LEGENDRE, O. EPMA Dating User's Manual, V.1.01: Age Calculation from Electron Probe Microanalyser Measurements of U-Th-Pb. BRGM, Orleans. 2004.

27. PYLE, J.M.; SPEAR, F.S.; WARK, D.A.; DANIEL, C.G.; STORM, L.C. Contributions to precision and accuracy of Chemical ages of monazite. American Mineralogist, v. 90, p. 547-577, 2005.

28. QUÉMÉNEUR, J.J.G. Archean and Trans-Amazonian magmatism in the Campos das Vertentes region, southern São Francisco Craton, based on geochemistry and geochronology. Institute of Geosciences, Federal University of Minas Gerais, Belo Horizonte, Full Professor Thesis, 79 p., 1995.

29. QUÉMÉNEUR J.J.G.; RIBEIRO, A.; TROUW, R.A.J.; PACIULLO, F.V.P.; HEILBRON, M. Geology of Folha Lavras. In: Companhia Mineradora de Minas Gerais. Geology and Mineral Resources of Southeastern Minas Gerais. South of Minas Project - Stage I. Final report. Belo Horizonte, COMIG/UFMG/UFRJ/UERJ, v. 1, p. 259-319. CD-ROM. 2003.

30. SCHERRER, N.C.; ENG, M.; GNOS, E.; JAKOB, V.; LIECHTI, A. Monazite analysis; from sarnple preparation to microprobe age dating and REE quantification. Schweizer Mineralogische und Petrographische Mitteilungen, v. 80, p. 93-105, 2000.

31. SMITH, H.A. & BARREIRO, B. Monazite U-Pb dating of staurolite grade metamorphism in pelitic schists. Contributions to Mineralogy and Petrology, v. 105, p. 602-615, 1990.

32. SUN, S.S. & MACDONOUGH, W.F. Chemical and isotopic systematics of oceanic basalts; implications for mantle composition and processes. In: Magmatism in the ocean basins. SAUNDERS, A.D. & NORRY, M.J. (Editors), Geological Society of London, London, v. 42, p. 313-345, 1989.

33. SUZUKI, K. & ADACFH, M. Precambrian provenance and Silurian metamorphism of the Tsubonosawa

paragneiss in the South Kitakami terrane, Northeast Japan, revealed by the Chemical Th-U-total Pb isochron ages of monazite, zircon, and xenotime. Geochemical Journal, v. 25, p. 357-376, 1991.

34. TEIXEIRA, W. & FIGUEIREDO, M.C.H. An outline of Early Proterozoic crustal evolution in the São Francisco region, Brazil: a review. Precambrian Research, v. 53, p. 1-22, 1991.

35. TEIXEIRA W.; CARNEIRO, M.A.; NOCE, C.M.; MACHADO, N.; SATO, K.; TAYLOR, P.N. Pb, Sr and Nd isotope constraints on the Archean evolution of gneissic-granitoid complexes in the Southern São Francisco Craton, Brazil. Precambrian Research, v. 78, p. 151 - 164, 1996.

36. TEIXEIRA, W.; SABATÉ, P.; BARBOSA, J.; NOCE, C.M.; CARNEIRO, M.A. Archean and Palaeoproterozoic tectonic evolution of the São Francisco Craton, Brazil. In: CORDANI, U.G.; MILANI, E.J.; THOMAZ-FILHO, A.; CAMPOS, D.A. (eds.) Tectonic evolution of South America. 31st International Geological Congress. Rio de Janeiro, p. 101-137, 2000.

37. TEIXEIRA, W.; ÁVILA, C.A.; CORDANI, U.G.; MARTINS, V.T.S.; VALENÇA, J.G. Isotopic data (U/Pb, Pb/Pb, Sm/Nd, Rb/Sr) from the Palaeoproterozoic plutonism of the Mineiro Belt, southern portion of the São Francisco Craton: tectonic implications. In: Symposium on the São Francisco Craton, 3, Salvador. Anais SBG, v. 1, p. 174-177, 2005.

38. VLACH, S.R.F. Th-U-PbrDating by Electron Probe Microanalysis, Part I. Monazite: Analytical Procedures and Data Treatment. Geologia USP - Série científica, v. 10, n. 1, p. 61-85, 2010.

39. WARD, C.D.; MCARTHUR, J.M.; WALSH, J.N. Rare earth element behaviour during evolution and alteration of the Dartmoor granite, SW England. Journal of Petrology, v. 33, p. 785-815, 1991.

40. WILLIAMS, M.L.; JERCINOVIC, M.J.; TERRY, M.P. Age mapping and dating of monazite on the electron microprobe: Deconvoluting multistage tectonic histories. Geology, v. 27, p. 1023-1026, 1999.

41. WILLIAMS, M.L.; JERCINOVIC, M.J.; HETHERINGTON, C.J. Microprobe monazite geochronology: understanding geologic processes by integrating composition and chronology. Ann Review Earth Planetary Sciences, v. 35, p. 137-175, 2007.

FINAL CONSIDERATIONS

In addition to non-Archaean mafic dykes, the Serra da Babilónia region is home to five distinct Neoarchean geological units, whose ages are summarised in Table I.

Table I. Compilation of ages for the units in the study area.

	Age (Ma)	**References**
Peridotitic Metaultramafic and Metanorites/Metagabronorites	2658 ±44	Pinese (1997)
Metagranitoid Capela São Francisco	Around 2700	Potassic granitoids from Romano *et al.* (2013)
MetachamockitoZMetachamoenderbito Rio do Amparo-Bom Sucesso	2718 ±13(1) 2708 ± 5 (2)	(1) Paciullo *et al.* (2007); (2) Quéméneur *et al.* (2003)
Metaenderbito/Metachranoenderbito Babylon	2730 (1) 2657 ± 36 (2)	(1) Quéméneur *et al.* (2003); (2) This work
Biotite-muscovite gneiss	Around 3,000	Carneiro *et al.* (2007)

Based on the field characteristics, the interpretation of aerogeophysical images, the petrography and geochemistry of the lithotypes investigated, the ages available in the literature and the age obtained in this study (Table I), and the geological model discussed in figure 22 (article 1), it is possible to suggest the following geological evolution for the different neoarchean units: by anatexis, the pre-existing biotite-muscovite gneiss in the lower crust of the southern sector of the São Francisco craton would have given rise to the peraluminous Capela São Francisco metagranitoid. Different degrees of partial melting of the previously metasomatised and enriched lithospheric mantle would have generated both the metalluminous Babylonian metaenderbites and metachamoenderbites as well as the metaultramafics. Through magmatic differentiation processes, these ultramafics would have evolved into metanorites and metagabronorites. The metalluminous metachamockites and metachamoenderbites of the Rio do Amparo-Bom Sucesso type were formed by hybridisation of material derived from mantle and crustal sources. Based on the above, the evolution of the Neoarchean magmatic and metamorphic processes in the Serra da Babilónia region would be related to the late to post-collisional events of the Rio das Velhas orogeny, within which there was the growth of phosphate-silicate crowns in the monazite crystals of the Babilónia unit, from the action of metamorphic hydrothermal fluids. All these geological processes took place under granulite to amphibolite facies conditions, corresponding to the lower crust between the Conrad and Mohorovicic seismic discontinuities, as shown in figure I.

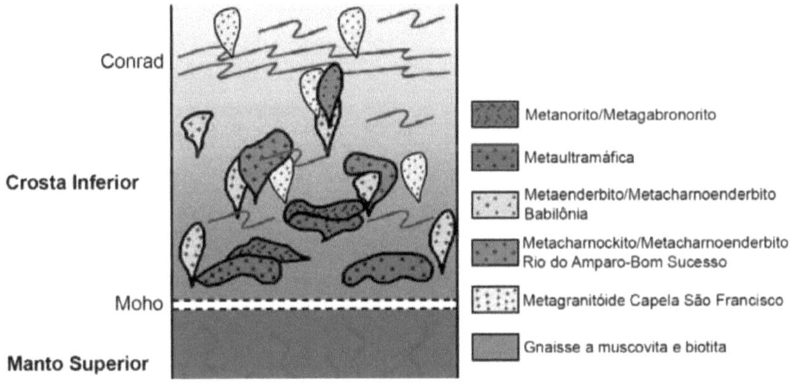

Figure I. Crustal position of the region's rocks during the Neoarchean.

BIBLIOGRAPHICAL REFERENCES

1. CARNEIRO, M.A. *et al.* Sheets Campo Belo (SF.23-V-B-VI) and Oliveira (SF.23-X-A-IV), scale 1:100.000: final report. Ouro Preto. Geology of Brazil Programme: Geological Maps. 192 p. 2007.

2. PACIULLO, F.V.P.; RIBEIRO, A.; CHERMAN, A.; CHRISPIM, S.; MACIEL, R.R. Geological map of the Nepomuceno Sheet (SF.23-V-D-III) Scale 1:100.000. Explanatory note. Geology of Brazil Programme: Geological Maps, Rio de Janeiro, 92 p., 2007.

3. PINESE, J.P.P. Geochemistry, Isotopic Geology and Petrological Aspects of the Precambrian Mafic Dikes of the Lavras Region (MG), Southern Portion of the São Francisco Craton. PhD Thesis, Inst. of Geosciences, University of São Paulo, São Paulo, 178 p., 1997.

4. QUÉMÉNEUR J.J.G.; RIBEIRO, A.; TROUW, R.A.J.; PACIULLO, F.V.P.; HEILBRON, M. Geology of Folha Lavras. In: Companhia Mineradora de Minas Gerais. Geology and Mineral Resources of Southeastern Minas Gerais. South of Minas Project - Stage I. Final report. Belo Horizonte, COMIG/UFMG/UFRJ/UERJ, v. 1, p. 259-319. CD-ROM. 2003.

5. ROMANO, R.; LANA, C.; ALKMIM, F.F.; STEVENS, G.; ARMSTRONG, R. Stabilisation of the Southern portion of the São Francisco Craton, SE Brazil, through a long-lived period of potassic magmatism. Precambrian Research 224, p. 143-159. 2013.

MIX
Papier aus verantwortungsvollen Quellen
Paper from responsible sources
FSC® C105338

Printed by Books on Demand GmbH, Norderstedt / Germany